WITHDRA

First published 1998 by
Liverpool University Press
PO Box 147, Liverpool L69 3BX

British Library Cataloguing in Publication Data
Data are available

ISBN 0 85323 068 4

Printed and bound by
Graphic Offset Ltd, 3-4 Phillips Street, Liverpool L3 6HZ

PRIMARY SPACE PROJECT
RESEARCH REPORT

1998

FORCES

by
**TERRY RUSSELL, LINDA McGUIGAN
and ADRIAN HUGHES**

LIVERPOOL UNIVERSITY PRESS

ACKNOWLEDGMENTS

The research reported here was made possible as the result of the dedication and professionalism of the staff of all the schools which participated. Our thanks go to them and the children whose ideas are the subject of this report.

To Lisa Caslin and Lucy Boardman, our thanks for their painstaking care and patient attention to detail with the DTP demands of producing this report.

CONTENTS

Contents

1. INTRODUCTION

1.0   Background to the project

This research report builds upon the methodology developed in the course of a significant programme of research by CRIPSAT over the last decade. There are several reasons for revisiting the domain of forces following some preliminary exploration by the SPACE project team in 1989. That Nuffield funded original research was conducted prior to the introduction of the National Curriculum. The original research informed the development of the curriculum materials published in the Nuffield Primary Science Series (CollinsEducational) but was not published in the form of a separate research report. The introduction of a prescribed curriculum carried with it a new set of demands. Firstly, there was a change from most teachers' emphasis, which was primarily a process approach, to a concern for both content and processes. Underpinning a nationally prescribed curriculum is a pupil entitlement to certain content and teachers' obligation to deliver that content. It was important that, in this new educational context, the domain of forces should be revisited in order that the research should have direct curriculum validity.

A second consideration was that the implementation of a prescribed curriculum across four Key Stages opened possibilities of the idea of progression in pupils' conceptual understanding being on the research agenda. Concerns about matters of progression cannot be confined to a single classroom or Key Stage. The 'Evaluation of the Implementation of National Curriculum Science KS1-3' project (Russell et al., 1994) demonstrated the insights which could be gained through exploring conceptual progression with teachers across the three Key Stages, KS1-3. Whereas the original SPACE research limited itself to the first two Key Stages, the experience of working with an extended age range made the advantages of this approach clear. Primarily, this involved a recognition of the importance of the concept of progression to a constructivist approach to teaching and learning science. Fundamental to this approach is the idea that, rather than being either right or wrong in their thinking, children's ideas are more usefully considered as revealing commonly recurring intermediate understandings. The greater the age range researched, the clearer the emergence and shifts in such ideas was apparent. Indeed, there is a strong argument in favour of extending any such research beyond the age range which is the immediate concern or focus for action, since it is only when antecedent and subsequent ideas are revealed that progression in the thinking of the age group under direct consideration are placed in developmental context.

A third consideration was that the implementation of a prescribed curriculum, together with its associated assessment arrangements, placed increased demands on teachers' own understanding of science. It has been argued that many primary teachers lack an appropriate level of understanding for teaching National Curriculum science. Forces in particular has been raised by teachers as a source of concern, (Kruger et al., 1990). Moreover, Russell et al. (1994; op cit.) suggest that limitations in teachers' understanding is not restricted to primary teachers but represents a source of concern for those Key Stage 3 and 4 teachers teaching outside their subject specialism. CRIPSAT's experience of helping teachers develop their understanding (Schilling *et al.*, 1992) was used to assemble some support materials for teachers involved in the Forces project. This material, together with the cross-Key Stage dialogue between teachers which was promoted by the nature of the research, attempted to support teachers in their pedagogical subject knowledge.

Another significant change in orientation, compared to the original SPACE project approach is that this research is underpinned by a particular formulation of conceptual progression which evolved from experiences gained in other empirical progression enquiries (Russell *et al.*, 1994). This formulation differs from attempts to describe cognitive development as something capable of being separated from socialised development, as Piaget attempted. Nor is our formulation of progression limited to a notion of an ideal, logical sequencing of ideas to be taught. Instead, it is explicitly recognised that these major factors - the individual's developing cognitive capacities and the particular set of ideas to which a society decides individuals will be exposed (or 'enculturated') - *interact*. Furthermore, we recognise that there are more effective and less effective ways in which teachers might support intellectual development. Teachers interpret the curriculum in slightly different ways and have their own methods of 'bridging' or 'scaffolding' between what is to be taught and the learners. Thus it is suggested that conceptual progression in an educational context might be understood as comprising these three interacting elements:

the sequencing of the curriculum agenda
developmental and experimental aspects of cognition, and
praxis, the teaching and learning strategies adopted by teachers.

Within an educational perspective, concentration on any one element alone would be insufficient; the three elements must be taken into account in any consideration of the nature of National Curriculum related conceptual progression. This notion of progression has led to a refinement of the mode of empirical enquiry (i.e. since the initial Forces research) to attempt to represent more fully the teacher's role in promoting conceptual understanding. Messages emanating from constructivist research approached from this standpoint promise to inform not just our knowledge of likely pathways of individual pupils' conceptual progression, but also pedagogical practice and optimal curriculum sequencing.

1.1   *The collaborative nature of the research*

The collaborative nature of the research reported here exploited the complementarity of skills offered by teachers and university researchers. The researchers brought with them a systematic research framework which had been assembled through the experience of earlier projects; teachers' ideas and practices were drawn upon and subjected to scrutiny by colleagues in other classrooms. Collaboration between teachers from three Key Stages provided opportunities to share ideas and practices and led to cross-fertilisation across traditional Key Stage divisions. Since this mode of enquiry is classroom-based it takes into account classroom realities (both constraints and opportunities) and examines what is possible within an ecologically valid context. Thus, it continues CRIPSAT's involvement in professionally-based enquiry underpinned by a collaborative methodology. The outcomes of such research have messages for classroom practice, as well as making a contribution to research understanding *per se*.

1.2   *Constructivist orientation*

The research is underpinned by a constructivist model of learning. Broadly stated, this view assumes that the learner is actively involved in constructing their own understanding of conventional science. The construction of these new ways of knowing are influenced by the ideas children have developed previously and the new ideas and experiences introduced by the teacher. That there is a socio-cultural dimension to these individual constructions is not neglected; personal constructions are influenced by the social context in which ideas are encountered. The culture, the teacher, the peer group and classroom situational effects are all influential in the formation of new understandings:

The university-based researchers' commitment to collaboration with teachers in research activity can also be described as constructivist in orientation. Ideas and practices were considered and developed within a series of group meetings with the expectation that the outcomes of these meetings would provide an agenda for action. Teachers made their own individual constructions of these events and operationalised the ideas and practices in their classrooms in their own ways. While the National Curriculum set the broad agenda for action, what might emerge in matters of detail was open-ended, benefiting from teachers' professionalism, expertise and imagination. Participation in the project and exposure to the views of colleagues expert in other Key Stages was no doubt stimulating. Teachers' interpretations of, and responses to, children's understandings demonstrate the fact that constructivist theory applies equally to teachers (and researchers) as to pupils. It is reflexive; conducting constructivist research is a constructivist activity. Good ideas, effective strategies for intervention, etc., were shared, discussed and recorded as useful outcomes of research activity.

1.3   *General approach*

The sequence of activities adopted for this research comprised:

Pre-intervention elicitation: a period during which teachers elicited the ideas of all children in their class using concept probes which addressed the concept domains being investigated.
Pre-intervention interviews: a stratified sub-sample of children were interviewed by university-based researchers.
Intervention: a period in which teachers helped children to develop their ideas through a range of empirical and other evidence-gathering activities.
Post-intervention elicitation: teachers collected children's ideas through a series of concept probes matching the pre-intervention probes (though not identical in kind or extent).
Post-intervention interviews: the same stratified sub-sample of children were interviewed by university-based researchers.

A detailed description of these phases is presented in Chapter Two. This research cycle is primarily constructivist in orientation. It was not intended to implement a tightly formulated experimental design in which pre- and post-tests are precisely matched, where commonly applied classroom interventions can be controlled and their effects measured. There are several reasons why these two elements of experimental design are inappropriate for classroom-based research. Constant comparison of data provides immediate feedback which assists in the further development of the data collection techniques. Insights are gained cumulatively and fed back into the programme *iteratively*. It is the nature of collaborative classroom based research that ideas and practices explored in the elicitation phase may be modified, refined or rejected at the point of post-intervention elicitation and interviews. Contexts which fail to engage children's interest may be replaced. It is the nature of the elicitation that ideas in a wide range of areas are probed. Post-intervention elicitation and interviews are tailored to examine changes in understanding in matters which have been approached within the intervention phase.

While a range of interventions are suggested, agreed and refined between teachers and researchers, it is important to recognise that teachers are not simply following an externally imposed input routine. Indeed, given increasing evidence of the role of social processes (Howe 1996) and context (Lave 1995) in promoting conceptual progression, evaluating the impact of specific routines has to be acknowledged to be intrinsically problematic. It is the nature of constructivist collaborative classroom research that teachers construct their own interpretation of the suggested activities and decide the nature and extent of intervention for a particular class. Teachers' judgements and practices are an integral part of the development and operationalisation of the intervention. They select and modify activities according to the prevailing ideas of children in their class and this makes it possible for them to monitor conceptual progression in the social context of classroom interactions. A balance has to be achieved between the need for rigour and systematic data collection and acceptance that the research process is situated in classrooms which are not sterile, controlled environments but are subject to myriad constraints and opportunities. Indeed, this is the very strength of collaborative classroom research, that messages emanating from such enquiry inform the development of theory as well as having classroom and curriculum validity.

Outcomes from such research have the potential to inform teaching and learning of science at Key Stages 1, 2 and 3. Valuable insights are also gained into aspects of the domain which would benefit from further enquiry.

The products of the research programme include the following:

- The report summarises the collaborative classroom-based methodology which informed the programme. It is assumed that this is a replicable methodology which has value in contributing to the development of researched curricula generally and more specifically, the process of review of the science curriculum in England and Wales.

- Much constructivist research itemises pupils' 'alternative conceptions'. We prefer the term, 'intermediate understandings' as this situates children's ideas in the socio-cultural context of science ideas while acknowledging the lack of complete correspondence between the two.

- As well as describing children's ideas, the report reviews qualitative and quantitative evidence of shifts in thinking. It is suggested that the programme could not optimise such developments since it was exploratory and iterative in nature. Nonetheless, certain areas are identified as being particularly promising targets for intervention.

- Given the definition of educational progression which guided this research as including the socio-cultural impact of teachers, it is logical that attention should focus on what appear to be promising strategies for helping children to develop their thinking in the direction of conventional understanding. Such possibilities are reviewed in the final chapter of this report.

- Although the National Curriculum is laid down in law as a pupil entitlement, we have treated it, for the purposes of the research reported here, as an hypothesis describing optimal content and sequencing. We have subjected that hypothesis to empirical enquiry alongside teachers in some of the classrooms in which it is required to be put into operation. This research was conducted during the period of the moratorium on change between 1995 and 2000, but it is anticipated that an important outcome will be a relevant contribution to the process of curriculum review.

- While educational research can never be complete so long as it assumes dynamic systems, we hope that the outcomes of this enquiry will contribute to a cumulative understanding of the science curriculum and teaching and learning science. Even within the domain of enquiry, there are obvious ommisions which will need to be the subject of further research.

- For the future, we anticipate a need to exploit and disseminate the findings reported here more widely through the medium of teacher support materials.

6

2. METHODOLOGY

2.0 Introduction

The sample, research programme and description of the project are described in sections 2.1. to 2.3.

2.1 Sample

a) Schools

Fifteen schools drawn from seven Local Education Authorities participated in this research. The schools are located in North West England and are situated in both urban and rural areas. Twelve are primary schools (Key Stages 1 and 2) while three are secondary schools (Key Stages 3 and 4). The classes included in the sample covered Key Stages 1-3 and ranged from Reception (usually aged five) to Year 9 (usually aged 14).

The names of the participating schools, Head Teachers and teachers are reported in Appendix One.

b) Teachers

Nineteen teachers were involved in the programme. Eight had previously been involved in collaborative research projects with CRIPSAT. Two had previously been involved in the SPACE project, one had participated in the Evaluation of the Implementation of Science in the National Curriculum project, (Russell *et al.*, 1994), five had participated with CRIPSAT staff in the development of end of Key Stage Two assessment materials. The other teachers had expressed a willingness to participate in the project. Many of these were working in schools in which the headteacher or other members of staff had been involved in collaborative work with CRIPSAT. Most teachers were thus aware of the constructivist philosophy which underpinned this project. Such was the enthusiasm of the teachers that at one school all of the three teachers within the small rural school were part of the sample, and in all cases were willing to modify their schemes of work to accommodate a Forces topic at a time convenient for the research program.

Teachers received support from the University researchers through three whole-group meetings. At the first of these meetings the teachers and researchers discussed and planned ways in which children's ideas could be found out. At the second, ways of helping children to develop their ideas were considered. At the third, changes in children's ideas, conceptual challenges and effective teaching strategies were discussed. Throughout the research period, visits to classes were made by the University researchers.

c) Children

All children in the fifteen classes were involved in the Project work to some extent. Pre- and Post-intervention concept probes were administered by the teachers to all children in the fifteen classes. Data were thus collected from 462 children. A stratified random sample of children was selected for interview at two stages during the work by University researchers. Teachers were asked to assign each child in their class to a Year group and an achievement band (high, medium or low) related to their overall school performance. The interview sample was then randomly selected from the class lists so that numbers were balanced by achievement band, Year group and gender. The sample comprised 118 children ranging from Reception to Year 9 and is summarised in Table 2.1.

Table 2.1   The Interview Sample

		KS1 n = 42	L KS2 n = 29	U KS2 n = 29	KS3 n = 18	
Low	Males	6	3	3	4	16
	Females	5	4	3	2	14
Medium	Males	6	8	8	4	26
	Females	12	3	5	2	22
High	Males	6	7	5	4	22
	Females	7	4	5	2	18
	64 Males 54 Females	42	29	29	18	118

2.2   The Research Programme

2.2.1   Review of associated research literature

There is a large accumulation of research literature addressing the domain of Forces. This literature documents children's ideas and has increasingly considered the possibilities for promoting conceptual progression within the domain. A comprehensive literature review was carried out prior to the commencement of this study. This review helped to inform the direction and starting points for this research. An outline review of the literature is presented in Chapter Three.

2.2.2   Introductory meeting of teachers and university-based researchers

An introductory pre-elicitation meeting was held in November 1996 prior to the commencement of the classroom based research. One of the aims of the meeting was to orientate teachers towards the nature of constructivist research. A second was to outline the research programme. A third consideration was the development of elicitation activities. Both previous research and the techniques developed during SPACE research influenced the selection of the range of possible activities presented to teachers. The suggestions were modified, clarified and added to as teachers discussed in groups whether and how the suggested activities might be tackled in their own classrooms. The range of activities were examined in terms of curriculum validity, Key Stage appropriateness and classroom manageability. The outcome of the meeting was an agenda for a sequence of conceptual probes. This was written up after the meeting and sent to teachers.

2.2.3   Classroom-based elicitation

Classroom work took place in two major phases. The first of these was called 'Elicitation'. Forces is a wide domain in which numerous concepts are embedded. Division of the domain into themes recognised the need to probe children's understanding in several areas. An attempt was made to achieve a compromise between the need for classroom-viable concept probes which would elicit the ideas of all children in numerous areas and the philosophy underpinning the research, that the elicitation of children's ideas should not involve ambushing children with decontextualised questions. Concept probes were thus developed which explored the areas of interest and could be presented within the context of a direct or familiar activity. The mode of response was primarily paper and pencil to maximise classroom viability. Two concept probes booklets were developed, one for Reception and Key Stage 1 and another for Key Stage 2 and 3. Concept probes loosely tailored to Key Stages ensured that children would be able to access and respond to the problems as posed. These are exemplified in Appendix II.

It was intended that teachers should introduce the concept probes with the presentation of associated direct experiences. Exploration of children's views of forces and their effects on objects in water, for example, was supported by a tank of water and a number of objects. Children were encouraged to engage with this direct experience. In the same way exploration of floating and sinking in air was supported by experience of a helium balloon. It was thought that direct experiences would stimulate some reflection on the domain and ensure more considered views were collected rather than catching children by surprise. Opportunities to draw on related everyday experiences were also fully exploited. Researchers distributed items such as helium-filled balloons, magnets and video material to each teacher.

The subject matter presented to children was selected on the grounds that it was likely to be familiar to them. The research team took account of the situated nature of cognition and concerns that children's ideas may be expected to be influenced by both the context and content of assessment materials. Since children's ideas were recognised as being both content- and context-sensitive, there was careful selection of both contexts and content in the development of concept probes. Awareness of context sensitivity resulted in some aspects of forces being explored through more than one situation.

The exact design of the conceptual probes was greatly influenced by knowledge and expertise gleaned from previous research. Many of the teachers and university-based research staff had become aware of the kinds of activities that had proved successful in meeting the dual functions of 'exposure' and 'elicitation'. It was anticipated that a wide range of aspects of forces could be explored with a large number of children in a manner that was classroom-viable. If teachers were to be enabled to explore the limits of children's understanding, such a range was essential.

The successful implementation of these activities as elicitation tools was based on techniques such as the use of open questions by the teacher in discussion with children. More fundamentally, teachers were aware of the rationale for adopting those techniques which enabled them to establish children's ideas. This required teachers to adopt a role in which they deliberately held back from guiding children's thinking during elicitation. They were to help children to clarify their ideas rather than seeking to make them justify or reconsider them. The latter form of activity would need to await the intervention phase.

2.2.4 Interviews

A sample of at least six children from each of the classes in the study was interviewed individually (n=118). The sample was balanced for gender and achievement and pupils were randomly selected within these constraints. Semi-structured interviews were used to explore further the ideas emerging from the concept probes. Key questions within the concept probes were identified as foci for the interview. However, in order to maintain informality, interviewers were encouraged to rephrase the questions where children were unclear about what was being asked. They also followed up children's lines of thought by asking additional questions in as spontaneous a manner as possible. Allowing for a flexible response to clarify as well as probe children's ideas meant that interviews ranged from about 20 minutes to one hour.

Members of the project team visited the schools and talked to children either in the classroom or in an otherwise unoccupied room. Permission to interview children was obtained from parents and/or teachers. The interviews were conducted in an informal manner and every attempt was made to put children as much at ease as possible. In fact, in several classes children were not only willing to express their ideas but also were extremely keen to do so.

2.2.5 Development of intervention activities

A second meeting was held with the participating teachers following the completion of the elicitation phase. The meeting was attended by the teachers and university-based researchers. Summaries of children's emerging ideas were considered and possibilities for intervention discussed. Micro-domains which were to become the foci of intervention and teaching and learning activities were explored and selected against the background of children's ideas and curriculum appropriateness.

2.2.6 Classroom-based Intervention

The second major classroom-based phase was termed 'intervention'. During the previous phase, teachers had been 'holding back' to encourage children to express their own ideas. In the intervention, teachers offered children experiences which gave them an opportunity to reflect on their ideas, test them out, discuss them and amend, reject or retain them. Several data sources provide evidence of the nature of intervention strategies adopted in each classroom. Teachers were encouraged to record the range of interventions used in their classrooms in journals. Where possible, they indicated how these strategies related to children's initial ideas and the outcomes of such interventions. Teachers were also encouraged to describe those aspects of intervention - such as a class discussion - for which a written report by a child would not be available. University-based researchers visited classrooms during the classroom intervention to collect examples of teaching and learning strategies operating in different classrooms. Field notes and video material were used to record the evidence collected during these visits. A further data source was the actual work completed by children and collected by teachers during the intervention phase. Teachers were asked to collect as much of children's classroom work as possible. Detailed descriptions of the intervention activities are presented in Chapter Five.

2.2.7   Post-Intervention elicitation of children's ideas

Intervention was followed by a post-intervention elicitation using concept probes in a similar manner to that described in the initial elicitation. These concept probes examined a narrower range of themes than in the elicitation and focused in particular on themes addressed during the intervention. Concept probe booklets were developed for children in each of the three Key Stages. Teachers for the most part administered the probes to the whole class. Individual interviews were carried out with the same stratified sub-sample as before (n=118).

2.2.8   Reflection and evaluation

A final meeting attended by teachers and university-based researchers was held following the completion of the classroom-based part of the research. At this meeting, shifts in children's ideas were discussed and obstacles or conceptual struggles which appeared to limit children's progress were identified. Particularly effective teaching strategies were shared. The value of collaborative action-research as a means of professional development was discussed. The limited scope of the intervention and targets for future research in this domain and others, were also recognised.

2.2.9   Data analysis and reporting

Interview responses, field notes of classroom visits, teacher meetings, video recordings and children's work provide a rich data source. Interview data were subjected to a long and detailed scrutiny in order to generate a qualitative set of exhaustive and mutually exclusive response categories. These coding frameworks were checked between team members and a coder was then trained. The interview sample's responses were coded, analysed and summarised. Field notes, children's work and impressions from teachers were the subject of constant comparison during the data collection process and informed the development of each phase of the research. Interpretation of the interview data was informed by some of these insights gained during the research programme.

2.3   Structuring the Research Domain

In the early stages of the SPACE project, a list of concepts was drawn up for each of the topics to be researched. This list was intended to delineate the boundaries within which the work would be carried out. With the advent of a National Curriculum for England and

Wales, the document for Science provided a framework in which concepts for investigation could be identified (DES 1993).

2.3.1   Division into Themes

The National Curriculum defines programmes of study (PoS) for different age groups of children. The programmes of study indicate the kinds of experiences to which all children of given age groups are expected to be exposed. That for Key Stage 1 (KS1) is applicable for children up to the school year in which they have their seventh birthday, that for Key Stage 2 (KS2) being applicable for children up to the school year in which they have their eleventh birthday, and that for Key Stage 3 (KS3) being applicable for those children up to the school year in which they have their fourteenth birthday.

Each subject of the National Curriculum is divided into a number of Attainment Targets (ATs). Each target is further subdivided into domains. At the time when this research was planned and carried out, there were four such targets designated for Science. The findings of this report pertain to the domain of Forces within 'AT4 *Physical Processes'* in the Science National Curriculum document

The researchers divided Forces into themes, each of which seemed to reflect a relatively discrete concept area within the domain. This process was informed by the National Curriculum *'Forces and Motion'* P.o.S. and the accumulated research literature which reports children's ideas in Forces.

A wide range of themes was addressed within the elicitation phase. This range was subsequently reduced following the elicitation of ideas. An examination of children's ideas across the themes revealed a range of alternative ideas. In one or two themes, children's understanding seemed to be at some distance from the target concept. In terms of conceptual hierarchies these themes were at the upper end of conceptual progression. It was considered to be more useful to focus on the more foundational themes. It was anticipated that making progress with understanding in these themes might provide a better basis for approaching other themes in subsequent learning episodes.

2.3.2   Issues Within the Focal Themes

The following issues were chosen from within themes to form the focus of the research.

a)   **Effects of forces**

What do children count as a force? What do children understand to be the effects of forces? How do children explain the behaviour of stationary and moving objects? What understanding do children have of non-contact forces?

b)   **Gravity**

What understanding do children have of gravity and its effects at a local level and on a planetary scale? To what extent are children able to distinguish between mass and weight?

c)   ***Friction and air resistance***

What understanding do children have of friction and its effects? How do children conceptualise air resistance?

d)   ***Reaction Forces***

What understanding do children have of reaction forces? Are they aware of such forces in both dynamic and static situations?

e)   ***Multiple Forces***

What understanding do children have that more than one force may simultaneously act on an object? Can children recognise and explain the different effects when forces are balanced or unbalanced?

The exact manner in which the compression and reduction of themes occurred will be discussed further in Chapter Four.

14

3. LITERATURE REVIEW

3.0 Introduction

This review broadly outlines research within the domain of forces, an area which has received much interest from constructivist researchers. An initial review of the available research revealed aspects of forces which had received most of researchers' attentions and also showed that research had tended to neglect the developing understanding of younger children. Interest has not focused exclusively on the accumulation of children's ideas but has also begun to consider the ways in which teaching-learning interactions might help promote understanding (Pfundt and Duit 1997). It is possible to discern differences in approaches to the elicitation of children's reasoning adopted by different researchers. The approach adopted by the SPACE project is distinct in many ways since an important aspect of the approach is the ecological validity of the ways of eliciting children's reasoning. That is, that the methods used to establish children's initial understanding are based on the non-directive rationale of the Piagetian individual clinical interview, with the important difference that the probes must be 'classroom viable' or capable of being used by teachers in their normal classroom practices.

The literature review is structured in three main sections:

3.1) summary of research approaches to the elicitation of children's ideas;
3.2) summary of children's ideas;
3.3) teaching strategies

3.1 Summary of research approaches to the elicitation of children's ideas.

The elicitation of children's ideas in schools raises many issues including the practical problems of organising and carrying it out, ethical issues such as gaining permission from school, parents and pupils and methodological issues. While the importance of management and ethical issues are recognised it is the methodological issues which are the focus of this chapter.

The individual clinical interview adopted by Piaget, (1929) has come to be recognised as a powerful tool in the elicitation of children's reasoning in science. Interviews enable researchers to gain access to an individual's understanding of events within different domains. Lythcott and Duschl, (1990) describe a distinction between the clinical interview which relies on verbal data and clinical exploration method which yields data based on what the child does. Current elicitation techniques adopted by researchers often include modifications to the classical clinical interview and often include data gathered using both approaches. Interviews enable the researcher to seek a more elaborated response and clarification of emerging ideas and are generally audio-recorded.

A favoured elicitation method with a long tradition is that of 'interview-about-instances', (Osborne and Gilbert, 1980, Watts, 1982). This technique probes children's understanding by posing open questions about particular instances or problems presented in pictorial form. A development of the interview about instances technique has been adopted by Bliss and Ogborn, 1994. They elicited children's reasoning about motion through comic strip sequences It was considered that comic strips had broad appeal to a wide age range and in addition, probed children's understanding of impossible events.

Another approach is the use of survey methods to gather children's written responses to questions posed in a questionnaire. Surveys tend to comprise closed question items and/or multiple choice items (Halloun and Hestenes, 1985; Kuiper, 1994; Palmer, 1997). Fischbein *et al.,* 1989 used a questionnaire supplemented by individual pupil interview to probe impetus conceptions. Open questions may be used within a pencil and paper format to explore understanding of particular events (Clement, 1982; Eckstein and Shemesh, 1993). Arnold *et al.,* (1995) asked children to represent their ideas about the Earth's shape and gravity in drawings, common features then being categorised and coded.

Encouraging children to handle objects and materials, or to draw on some direct experience is increasingly used to encourage the elicitation of children's ideas (Bar, *et al.,* 1994). Gair and Stancliffe (1988), claimed that encouraging the manipulation of objects provided a non-threatening environment for the expression of ideas. Concerns that researchers' questions might steer children towards particular responses led Hamilton, (1996) to explore children's ideas through peer interview. A paired-interview approach was also adopted by Millar and Kragh, (1994) in their examination of children's explanations of the motion of projectiles.

The approach adopted by the SPACE project is to collaborate with teachers in the elicitation of children's understandings. Teachers encourage the free expression of ideas by posing open questions. The atmosphere created in the classroom is one in which children's ideas are valued as being provisional. The ecological validity of the manner in which children's ideas are elicited is the foundation for researchers' confidence in knowledge claims. Ecological validity is achieved through the development of diagnostic concept probes administered by teachers as an extension of their classroom techniques. An important aspect of such concept probes is that children are not ambushed by a decontextualised challenge. Their thinking is orientated towards the domain through the provision of a practical or familiar experience. Children are encouraged to respond in writing, drawing or orally. Confidence in the knowledge claims associated with data collected in this manner (Lythcott and Duschl, 1990) is further reinforced as university researchers carry out audio-taped individual interviews, with a sub-sample of children from each class. These interviews involve probing further ideas expressed through the concept probes. Researchers display unconditional positive regard for children's expressed ideas. They carefully probe emerging ideas and hesitations in order to gain additional insights into children's understandings. It is argued that the collaborative nature of the research means that that the techniques and sequences adopted within the research correspond to a constructivist teaching and learning sequence (McGuigan and Russell, 1997). The gap between theory and practice is narrowed; research data are centred on an agenda of what is possible in classrooms.

3.2   *Summary of children's ideas*

The domain of forces has received intense interest from the research community. The accumulated data contrast sharply with the dearth of evidence on pupils' ideas about, for instance, sound. This section briefly outlines the scope of research focusing where possible, on the age groups of the sample in the study presented here. Interested readers are urged to consult the primary sources for more detail. The presentation of the literature review will be structured according the themes adopted in the research reported in this volume. The research documents the range and commonalties in children's understanding and the different ways in which children explain the effects of forces. The situated nature of cognition, is illustrated; researchers frequently report that children do not appear to have a consistently applied set of beliefs about forces and their effects. Their beliefs about individual forces appear to be influenced by the specific instance in which the force is operating as well as the nature of the problem posed (Kuiper, 1994).

Recognition of the context-specific nature of children's understanding is not new (Wason, 1977; Donaldson, 1978). The situated nature of cognition has been more recently examined by Lave, (1995) and Hennessy, (1993). In science, examples of how children's understanding of forces differs in different contexts have been reported in the literature. Halloun and Hestenes, (1985) report that university students were inconsistent in their application of concepts across contexts. Palmer, (1997) reported that 15-16 year olds were more influenced by contexts than pre-service teachers. Engel-Clough and Driver, (1986) suggest that it is in those contexts which, according to the children's frame of reference seem to explore different phenomena that inconsistency in conceptual frameworks are revealed. Reif, (1987) suggests that students tend to invoke 'knowledge fragments' rather than generalised understandings in response to physics problems which leads them to express inconsistent ideas. The suggestion that children tend to apply different ideas in different contexts casts doubt, according to Kuiper, (1994), on suggestions that children operate with coherent but different theories to those advanced in science. Kuiper challenges the idea that children operate with alternative frameworks and suggests instead that children could be understood as operating with a loose set of incoherent ideas. Di Sessa, (1989) suggests children's reasoning is fragmentary, the term 'phenomenological primitives' is used to describe the nature of children's ideas. Millar and Kragh, (1994) attempt to identify some of the factors influencing the use of different reasoning in different contexts. They suggest that familiarity of the context, the weight of the projectile and the assumed action of the wind all influence children's causal explanations about the trajectory followed by a passively dropped projectile. Palmer, (1997) found 15-16 year olds' reasoning to be similarly influenced by context. Whitelock (1991) reports that amongst 7- 16 year olds, the active nature of animate moving objects was a powerful influence on causal explanations.

Other theorists suggest that, while children appear to operate with competing and inconsistent theories, there *are* coherent frameworks underlying their reasoning in the forces domain, (McCloskey, 1983; Ogborn, 1985; Bar *et al.,* 1994; Bliss *et al.,* 1989; Mariani and Ogborn 1991; Gutierrez and Ogborn, 1992; Bliss and Ogborn 1993).

Much of the research on ideas about forces has, in the past, been conducted with adolescents and college students. More recently there has been increased, though less extensive, interest in the conceptualisations of younger children. This research review summarises evidence of students' ideas ranging from four years to adulthood. It is of interest that, while some studies report age-related differences in the application of causal explanations across contexts, (Palmer, 1997) many studies report little age-related progression in causal explanations about different aspects of forces (Andersson, 1990; Sequiera and Leite, 1991; Brown and Clement, 1992; Twigger *et al.*, 1994; McDermott 1984).

3.2.1 Gravitational Force

It is suggested that children tend to conceptualise gravitation as a force which requires a medium through which to travel (Andersson, 1990; Bar *et al.*, 1997). The medium most commonly cited as necessary for gravity is air. Sometimes air is conceptualised as the atmosphere in its entirety, as a single-entity, macroscopic view. Others suggest the single particles which comprise air transmit gravity from one to another (Watts, 1982 in a study with pupils in the 6-12 age range). Belief in the necessity of air as a medium for gravity leads to a number of associated understandings. A corollary of this idea is that where there is no air such as in space, there is no gravity (Watts, 1982; Stead and Osborne, 1980; Osborne *et al.*, 1981; Twigger *et al.*, 1994). Another view associated with the need for air is that there is less or even no gravity in water (Osborne *et al.*, 1981; Stead and Osborne, 1980).

Watts (1982) suggests that pupils in the 12-18 age range regard gravity as a constant and that moving objects try and fail to counteract its effects. McCloskey (1983) reports some understanding in his students that gravity can act as a ball moves upwards. Students' understanding of the trajectory of the ball included a view that, as the ball moves upwards the impetus force dissipates and gravity takes over; as the ball peaks, gravity and impetus force are regarded as equal; as the ball falls, gravity takes over. Indeed, Clement (1982) found that only one third of university physics students correctly represented gravity as the only force other than air resistance on an object moving upwards. The situated nature of children's understanding of gravity is suggested by the seemingly inconsistent ideas held. Many children regard gravity as selective in its operation and effects. There is a tendency to explain that it doesn't act on all things in the same way or even on the same things in the same way at all times (Watts, 1982). Items frequently associated with having gravity are heavy, inert things such as heavy boots. Lively, active objects are more likely to be assumed to counteract gravity. The speed of an object is often found to be assumed to influence the amount of gravity operating upon it. Watts (1982) found that the effects of gravity are believed to be greater when an object moves more slowly.

Some research point to a lack of awareness (or use) of the concept of gravitational force in explaining objects motion. Gilbert *et al.*, (1982), in a study of students between 7 and 20 years, reported a failure to suggest gravity acting on a ball moving downwards in favour of the idea that the ball moved freely. This is consistent with Piaget (1929) who also recorded children not using gravity in their explanations. The absence of gravity in some responses may be due to a widely held view that unsupported objects fall (Ogborn, 1985; Bliss, 1989; Vosnaidou and Brewer, 1992 and 1994; Eckstein and Shemesh, 1993). Reyneso *et al.*, (1993) found that eight and nine year olds found it sufficient to reason that objects fall 'downwards' while ten year olds suggested that objects fall to Earth because of gravity and float on the moon because of an absence of gravity. Gravity tends to be associated with downward motion. It is described as a force that 'pulls things down' or something that 'keeps objects down' or something that 'stops them floating away' (Bliss *et al.,* 1989; Bar *et al.*, 1997). No pupils in Bar's sample of 9 to 18 year olds expressed a view that gravitational forces act between two masses. There is evidence that some pupils consider it to be an upwards force and that it can be conceptualised as a push (Osborne *et al.,* 1981). A study Watts (1982) identified the belief that gravity begins to act when an object is falling towards the ground and stops acting when an object comes to rest on the ground. (Participants in Watts' study claimed that a golf ball was acted upon by gravity as it was falling but that gravity ceased acting when the golf ball came to rest). In contrast, Twigger *et al.,* (1994) report that 88 per cent of children aged 10-15 years suggested that gravity operated all the time. (There may be some changes in the pattern of responses towards an increased awareness of gravity as a result of National Curriculum in England).

Bar *et al., (op.cit.)* report an age-related trend in explanations of why objects fall. Young children hold the view that objects fall because they are *unsupported* (4-7 years). This idea persists and combines with causal explanations that the *heaviness* of objects causes them to fall (7-9 years) and later with an emergent idea of the Earth's attractive force (9-13 years). Arnold *et al.,* (1995) probed children's understanding of the Earth's shape and gravity (7-11years). They found that the majority (96 per cent) of children recognised the Earth as spherical but while they reasoned that gravity acted 'downwards' they failed to demonstrate any understanding that gravity acts towards the centre of the Earth.

Quantification of forces is an important aspect of developing understanding. While there is a paucity of evidence about children's understanding of quantification of forces, there is some evidence (Watts 1982) that children understand gravity to be a large force. According to Watts, gravity is conceptualised as a large force because it keeps so many objects on the ground. Children fail to appreciate that the size of gravity is related to the quantity of matter of two objects and the distance between them. A commonly expressed view is that at a local level rather than at an astronomical level, the magnitude of gravitational force changes according to changes in the height of the object above the ground. Furthermore, in this view, gravity increases as an object moves higher (Watts and Zylberstajn, 1981). Those holding this view might suggest that objects higher up would need more force to support them. For instance, a car situated higher up a hill would require more force to support it (Watts and Zylberstajn, 1981; Watts, 1982; Osborne *et al.,* 1981). Suggestions that gravitational force increases as objects move further apart are contrary to Newton's law that gravitational force decreases as the distance between objects increases. Conventional explanations usually treat gravity at a local level (i.e. on the surface of the Earth) as a constant force, while children often treat is as variable.

The science community is equivocal as to whether gravity and weight should be conceptualised as different phenomena. The accumulating research evidence presents a picture of children's understanding of the relationships between gravity, weight and mass. Children tend to distinguish between gravity and weight (Ruggiero *et al.,* 1985). They tend to believe gravity is not closely connected with weight but acts *with* weight to hold things down. Osborne *et al.,* (1981) examined the understanding of students between nine years and tertiary level, some of whom expressed the belief that it is possible to have weight without gravity. Weight is an attribute of an object (Ruggiero *op.cit.*) while gravity is associated with *movement downwards.* (Watts, 1982). Ruggiero *et al., (op.cit.*) explored the ideas of children in the age range 12-13 years about falling objects. They report three kinds of explanations. Firstly recognition of the role of gravity which operates on the weight of objects causing them to fall. Secondly, suggestions that weight and gravity *independently* cause objects to fall. Thirdly, explanations in which fall is due to the absence of support, and where weight and gravity are kept separate. Galili (1993) and Galili and Kaplan (1996) argue that it might be useful to encourage children to define gravity and weight as separate concepts. They suggest an approach which encourages a definition of weight as a force against a support, which can be distinguished from gravity. Bar *et al.,* (1994) probed children's (4-13 years) understanding of weight. Some of the everyday understandings of heaviness they gathered are consistent with other research, including an idea that heavy things exert a force on supporting objects (Minstrell, 1982).

Further confusions emerge when children's appreciation of the distinction between mass and weight are probed. Mullet (1990) records that, regardless of age, *weight* and *mass* tend to be conceptualised as weight, whereas, the *amount of matter* a thing is made of is related to the concept of *mass.* According to Mullet, difficulties in appreciating the distinction between mass and weight are exacerbated by everyday conceptualisations of weight and a tendency to use weight and mass interchangeably It is argued that weight is used in everyday operations when mass is being considered. Furthermore, in an everyday sense, mass can have several meanings - a mass of people or a church service. The substitution of the phrase *amount of matter* for the term mass seems to be of some value in helping students conceptualise the distinction between mass and weight.

3.2.2 *Friction*

Osborne *et al.,* (1981) examined children's ideas about friction using illustrated cards and interviews about these instances. They found that more than half of children in their sample associated friction with rubbing two surfaces together. As a consequence, friction was commonly associated with heat energy. A second group of ideas was to associate friction with movement. From this view point, no movement meant no friction. A third view expressed by a few students was that friction *was* the movement. A fourth view tended to consider friction as a force. Some responses suggested that friction is a force between two objects. Friction tended to be associated with solid objects although a limited number of children believed friction could occur with liquids. The possibility of friction occurring in air was often rejected.

Twigger *et al.,* (1994) explored children's understanding (in the age range 10-15 years) of friction through several hands-on tasks coupled with interview. The idea of friction seemed to feature in children's explanations at around 11 to 12 years. Their recognition of the role of friction in slowing movement of objects varied according to the context of the task. In the context of brakes being applied to a cycle only 31 per cent mentioned friction as a force. In contrast, in the context of a child kicking a pebble, 61 per cent referred to friction. When interviewed about a railway carriage slowing down or coming to a stop 94 per cent suggested friction, or air resistance (42 per cent). In the context of riding a bicycle 85 per cent mentioned air resistance as a opposing force and 77 per cent mentioned gravity. The increased mention of air resistance in the context of riding a bicycle was thought to be associated with personal experience of air resistance when riding. Detailed probing of children's ideas suggested to Twigger *et al.*, that, in most cases, there was a suggestion that even without friction or air resistance, objects would still come to a stop because the push would be used up. Even when children mentioned friction as a force they tended not to regard it as a force opposing motion. This latter view was reported by Stead and Osborne (1980).

3.2.3   Balanced and unbalanced forces

Twigger *et al.*, (1994) explored children's understanding of forces in instances when objects were moving at steady speeds and when they were accelerating. They found children tended not to recognise the relevant forces acting and often included additional irrelevant forces. Children were largely unaware of the correct relationship between forces. They found a widespread assumption that a constant force was needed to maintain motion accompanied by a belief that the forwards or driving force must be *greater* than any resistance in order for the object to be moving. Ninety-seven per cent thought that if these forces were equal then the object would stop. There was no age-related trend apparent in these explanations. A further task probed children's predictions of what would happen to a wheeled vehicle (a cart) being pulled with a constant force. All the students predicted that the cart would move at a constant speed. Over one quarter of the sample could not explain their observations of the accelerating cart, over half disbelieved that an elastic band was exerting a constant pulling force and incomplete explanations were offered by almost one fifth of children. Similar results were obtained in the context of riding a bicycle on level ground. Nearly all children equated a constant force applied by pedalling with constant speed. Most believed that if the forces were equal the bike would stop. Children's ideas about unbalanced forces were probed by asking them to consider the forces involved when the movement of an object was steadily increasing. All the children explained that a steadily increasing force would need to be applied. Thijs (1992) in a summary of children's ideas about motion, suggested that students' tendency to associate continuing movement with continuing force leads to several associated ideas. One of these is the belief that an *increasing* velocity needs an *increasing* force applied in the same direction as the movement.

3.2.4   Reaction forces.

Erickson and Hobbs (1978) found a tendency amongst a sample of 13-14 year olds not to appreciate that forces act in pairs and as a consequence, a lack of appreciation of reaction forces. Contrary to conventional science explanations, Brown and Clement (1987) report that High school children's reasoning about a collision between two objects led them to assume that the force on each of the two colliding objects was not equal. It was often assumed that the impact of the faster moving of two objects in collision would be greater. The idea of two forces interacting with each other is made difficult because of the language of *reaction* or *opposite* in the context of paired forces. According to Terry *et al.*, (1985) reaction conveys a sequence of events in which one force leads to a second. The word *opposite,* it is argued, suggests to some pupils that the action and reaction forces both act on the same object, simultaneously.

The idea that a push can be exerted by an inanimate, solid object like a table or a chair is according to Minstrell (1982) counter-intuitive. He found students invoked gravity as the explanatory concept to explain how a book is kept down on the table. Some students suggested that air or air pressure helped gravity. Only about half of these high school physics students suggested that the table might exert an upwards push on the book. Of those who cited a combination of forces, most believed that the downward push must be greater than the upward push in order to keep the book down. Simon *et al.,* (1994) noted a similar range of responses when they posed a similar problem with children citing single forces either down or up and occasionally suggesting a combination of forces. Other research has recorded children's tendency to regard gravity as the only force operating when objects are at rest (Kuiper, 1994; Twigger *et al.,* 1994). Gilbert and Watts (1983) identify a conceptual framework that 'no forces act on an object at rest'. However, Finegold and Gorsky (1991) found limited support for suggestions that no forces act on objects at rest. Only nine per cent of students consistently applied this belief across contexts and following teaching, the belief was no longer elicited.

3.2.5   Forces and their effects

Children's understanding of force is often associated with everyday usage of the term 'force'. For instance, everyday interpretations lead children to associate force with compulsion. As a result children might include the idea of being 'forced' to do something in their definitions (Osborne *et al.,* 1981; Gilbert *et al.,* 1982; Gair and Stancliffe, 1988). Gair and Stancliffe propose three frameworks in which children at age 11 years understand forces, ranging from an everyday interpretation of forces to an appreciation of forces as pushes and pulls.

Several researchers report the prevalence of a belief that movement implies a force and that an object requires a constant force to sustain its movement. (Galili and Bar 1992, with subjects aged 15 to adult, Bliss *et al.,* 1994; Champagne *et al.,* 1980; Osborne 1981 with 15 year olds; Watts, 1983; Enderstein and Spargo, 1996; Palmer, 1997) Researchers document a widespread belief in the idea of *impetus* as a force which keeps an object moving until it runs out or is used up (Schollum *et al.,* 1981; Watts and Zylberstajn, 1981; Clement, 1982). It has been suggested that the idea of an 'impetus' has parallels in the history of science, for instance, in the pre-Newtonian ideas of Aristotle (Champagne *et al.,* 1980) McCloskey (1983b,1988) argues that there is a distinction between Aristotle's ideas of an external force such as air which keeps an object moving and the later impetus theorists who believed an internal force was imparted to an object as it was released. Eckstein and Kozhevnikov (1997) conducted a large study of children in grades 3 to 12 and were able to discern three groups of responses which parallel historical ideas.

Two thirds of high school physics students believed a force is transferred from one object to another during impact (Steinberg *et al.,* 1990). An object is understood to acquire a force as it is released or projected and this force dissipates as the object moves along (Viennot, 1979; McCloskey *et al.,* 1983a, 1983b, 1988; Gilbert *et al.,* 1982). McCloskey suggests that the idea of an impetus force is associated with early experiences of objects starting moving and slowing down. Galili and Bar (1992) note the instability of conceptual change indicated by the regression to impetus theories following instruction. Indeed, research indicates the enduring nature of beliefs that a force continues to act in the direction of movement. It seems that this view prevails even amongst large numbers of students following physics instruction at the tertiary level (Palmer, 1997; Clement, 1982).

3.2.6 *Momentum*

Eckstein and Shemesh (1989) examined 9-16 year old students' reasoning about momentum. They found no age-related differences in children's causal explanations of where the ball released from a moving cart would fall. Children tended not to appreciate the continued forward movement of the ball, suggesting instead the ball would move straight down. Analysis of casual explanations revealed both intuitive and logical reasoning which was consistent from grade to grade. Fischbein *et al.,* (1989) noted that tenth graders gave 'straight down' responses for an object released by a moving carrier. Fischbein suggests that students are more willing to suggest continued forward movement in an *active* motion instance than in instances of *passive* motion.

Millar and Kragh (1994) investigated 11 year old children's understanding of inertia. They found that explanations of the movement of a passively thrown projectile from a moving car rarely indicated that the projectile continued its forward movement. On the other hand, there was some preparedness to suggest forward movement of a paper dart (not thrown, but released by a runner). Graham and Berry (1996) found only 14 per cent of students aged 16-19 years demonstrated a good understanding of momentum. McCloskey (1983a, 1983b) and McCloskey *et al* (1983) found only 45 per cent of college students reasoned that a passively released ball would continue its forward movement once it had left the hand.

3.3 *Teaching strategies*

The difficulties in helping children develop their ideas towards those more consistent with conventional science explanations is well documented (Gilbert *et al.,* 1982; Thijs and Van Den Berg, 1994). Bliss *et al.,* (1994), recognising the counter-intuitive nature of science explanations of forces, argue that striving for replacement of children's reasoning with a Newtonian view of force and motion is problematic because children's reasoning are developed from very early intuitions about causes of movement, yet in Newtonian physics, constant motion does not have a cause or require effort. Despite these difficulties there has been some examination of how children's understanding might be enhanced. Some studies report successful attempts to promote conceptual development in forces in a classroom setting. Some have explored particular teaching strategies whereas others suggest particular curriculum sequences. A few make recommendations for practice based on evidence of children's difficulties. Scott *et al.,* (1983) review teaching strategies within a constructivist approach and suggest that the range of approaches can be divided into two broad groups: those that focus on provoking cognitive conflict and those which take children's ideas and aim to encourage some development in those ideas. Schollum *et al.,* (1981) present a teaching programme which takes children's ideas as a starting point. This teaching-learning sequence comprises direct experiences and worksheets in which children's understanding is probed. The programme is structured so that firstly, children and their teacher became aware of the prevailing ideas in the classroom. A series of tasks are then suggested which aim to help children develop some key ideas. The first builds on children's intuitive idea of an impetus 'force' by labelling what is believed to be a force as 'momentum'. Further examination of the difference between force and momentum is followed by exploration of the effects of forces in combination. There have been other attempts to derive teaching sequences as a result of research into conceptions (Thijs, 1992; Dykstra *et al.,* 1992; Minstrell, 1982; Millar and Kragh, 1994).

Palmer (1997) suggests that teachers should endeavour to examine with children the different contexts in which a particular instance of a concept applies. In this way, children might be encouraged to generalise their understanding of a concept to new, unfamiliar instances in which the concept occurs. Champagne *et al.,* (1985) recommend provoking cognitive conflict by introducing discrepant ideas and events which can encourage reconsideration of existing conceptions and the construction of new understandings. Dykstra (1991) agrees with the need to provide children with events or ideas which challenge their existing ideas. According to this view, children should be encouraged to question their beliefs and recognise the inconsistency between competing explanations. Dykstra points to the limitations of provoking cognitive conflict and then adopting a transmission model of teaching. He describes how, once disequilibration is stimulated, teaching needs to be sensitive to the individual construction of meaning by encouraging children to put their own ideas to the test.

A teaching strategy which has attracted a large amount of research interest has been the role of analogies in the development of pupils' understanding. (Brown, 1994; Dagher, 1994; Clement *et al.,* 1989). The debate has centred on the nature of analogies and their use in the classroom. The contribution of analogies to children's learning is not straight-forward. Duit (1991) describes how analogies might help students make links between abstract concepts and the real world. However, children may make unintended links and may not recognise the limitations of an analogy. Treagust *et al.,* (1992) examined the practice of teachers who indicated analogies were a frequently used part of their teaching repertoire. They report that analogies were effective in promoting understanding when teachers used them as part of a well prepared teaching repertoire and when there was a teacher expectation that children would construct an understanding rather than passively receive the teacher's analogy. Bridging analogies which attempt to provide children with connections between existing beliefs and the target instance of a concept have been explored, Brown and Clement (1992). Brown (1994) examined closely pupils' interactions with analogies designed to help them appreciate the upward force of a table on a book. They found that analogous instances helped children towards an appreciation that the table exerts an upward force on the book. Analogies were found to be less helpful in helping children towards the abstraction that the upward force balances the downward force of the book. One of the outcomes of research, by Thijs and Bosch (1995) was that bridging analogies were found out to be successful in helping children to develop their ideas in the direction of conventional scientific understanding.

The exploitation of direct experiences has been readily incorporated into teachers' repertoires of practice and has recently attracted some research interest. Hatano and Inagaki (1992) suggest direct experiences make a positive contribution to children's learning. Thijs (1992) reports some changes in pupils' conceptions about motion as a result of direct experiences and opportunities for discussion. The link between direct experiences and discussion is an important one, for it ensures that children are not simply engaged in the physical manipulation of materials but are also encouraged to engage intellectually with the evidence of concepts informed by direct experiences.

Encouraging metacognition has increasingly being advocated as a mode of promoting conceptual change. Researchers (White and Mitchell, 1994; Kuhn, (1995) report that students can be encouraged, during discussion, to consider the content of their knowledge and the processes or strategies they employ in order to gain that knowledge.

In 1985, Gilbert and Zylberstajn suggested that constructivist teaching should make reference to the historical development of science ideas, especially in those areas such as impetus theories for which parallels could be found between children's ideas and the ideas held by conventional scientists. Reference to the historical development of ideas is also recommended as one approach by McCloskey and Kargon (1988).

26

4. ESTABLISHING CHILDREN'S IDEAS

4.0   Introduction

Previously published reports in this series have documented pupils' ideas prior to intervention. This report differs in this regard, the frequencies with which ideas occurred prior to and subsequent to intervention being reported in Chapter Six. There are several reasons for this difference in reporting structure. Primarily, the large volume of data collected across a broad range of sub-domains has made selection of a sub-set of results for publication essential. This chapter describes the National Curriculum agenda which was the starting point (though not a limiting factor). It also outlines the manner in which the concept probes were structured. The outcomes of the programme of research are presented in Chapter Six, following a review of intervention activities in Chapter Five.

4.1   Scientific content

As indicated in Chapter One the introduction of a National Curriculum had significant implications for teachers across all domains of science. Many felt that they were inadequately prepared for the new demands being made. Eight years later and the domain of forces in particular continues to give rise to considerable concern. The Programmes of Study headed 'Forces and motion' in the three Key Stages currently are as follows:-

Pupils should be taught:

at Key Stage 1:
   a) *to describe the movement of familiar things;*
   b) *that both pushes and pulls are examples of forces;*
   c) *that forces can make things speed up, slow down or change direction;*
   d) *that forces can change the shape of objects.*

at Key Stage 2:
   a) *that there are forces of attraction and repulsion between magnets, and forces of attraction between magnets and magnetic materials;*
   b) *that objects have weight because of the gravitational attraction between them and the Earth;*
   c) *about friction, including air resistance, as a force which slows moving objects;*
   d) *that when springs and elastic bands are stretched they exert a force on what ever is stretching them;*
   e) *that when springs are compressed they exert a force on whatever is compressing them;*
   f) *that forces act in particular directions;*
   g) *that forces on an object can balance, and that when this happens an object at rest stays still;*

> h)   that unbalanced forces can make things speed up, slow down or change direction.

at Key Stage 3:
>   a)   how to determine the speed of a moving object;
>   b)   the quantitative relationship between speed, distance and time;
>   c)   that unbalanced forces change the speed and/or direction of moving objects;
>   d)   that balanced forces produce no change in the movement of an object;
>   e)   ways in which frictional forces, including air resistance, affect motion;
>   f)   that forces can cause objects to turn about a pivot;
>   g)   the principle of moments and its application to situations involving one pivot;
>   h)   the quantitative relationship between the force acting normally per unit area on a surface and the pressure on that surface;
>   g)   some applications of this relationship.

For the most part these Programmes of Study delimited the range and depth of the research reported in this volume. However, the apparent assertion implicit in the programmes that a particular concept should be taught at a particular Key Stage, and not earlier, did not necessarily exclude that concept from the research with younger children. It was considered that, in some instances, attempts to gain a knowledge of children's earliest conceptions would be worthwhile because it would consolidate understanding of progression in their thinking.

Additional insights into the National Curriculum's assumptions about progression within the Forces domain can be gained from the formal descriptions of the 'Levels' of pupil attainment. The relevant extracts are:

Level 1 - Pupils describe the changes in movement which result from actions such as pushing and pulling objects.

Level 2 - They compare the movement of different objects in terms of speed or direction.

Level 3 - Pupils use their knowledge and understanding to link cause and effect in simple explanations of physical phenomena, such as the direction or speed of movement of an object changing because of a force applied to it.

Level 4 - They make generalisations about physical phenomena, such as motion being affected by forces, including gravitational attraction, magnetic attraction and friction.

Level 5 - They begin to use some abstract ideas in descriptions, such as forces being balanced when an object is stationary.

Level 6 - They use abstract ideas in descriptions and explanations, such as the sum of several forces determining changes in the direction or the speed of movement of an object.

Level 7 - They use some quantitative definitions, such as those for speed or pressure, and perform calculations involving physical quantities, using the correct units. They apply abstract ideas in explanations of a range of physical phenomena, such as the role of gravitational attraction in determining the motion of bodies in the solar system.

Level 8 - They use quantitative relationships between physical quantities in calculations that may involve more than one step. They offer detailed and sometimes quantitative interpretations of graphs, such as speed-time graphs. They use their knowledge of physical processes to explain patterns that they find.

It is clear that these Level descriptions do not cover comprehensively the Forces content of the Programmes of Study. The possible progressions in pupils' understanding within each of the strands of this domain are, therefore, in need of exploration by both teachers and researchers.

Additionally, it was not thought prudent to be totally constrained by the content of the current National Curriculum as if it were the last word on the subject. In consequence, consideration was also given to other problematic aspects of 'Forces' gleaned from the literature review outlined in the previous chapter. At the initial exploration phase the net was deliberately cast wide in order to maximise the possibility of capturing insights into children's formative understandings across a broad sweep of the Forces domain.

With this principle in mind the activities and associated questioning covered children's understanding of the following areas.

Vocabulary
the scientific meaning of the word 'force'.

Force and Motion
changes in movement necessitating a force;
static situations requiring the forces to be balanced;
unbalanced forces resulting in acceleration;
constant speed requiring the forces to be balanced;
the qualitative relationship between force, mass and movement.

Friction and air resistance
the effects of friction and air resistance on movement;
the effect of speed on air resistance.

Deformation effects
the deformation effects of forces.

Forces which act at a distance
forces which act without contact - magnetic, electrical and gravitational;
the effect of distance on these non-contact forces;
the nature of gravity;
the effects of gravity on the movement of objects;
the concepts of mass and weight.

Reaction forces
reaction forces.

Upthrust, density and pressure
upthrust in liquids and gases;
the link between upthrust and density and pressure differences.

Measuring forces
methods and units of the measurement of forces.

Arrow conventions
the conventional representation of forces using arrows.

Pressure
the concept of pressure.

Turning effects
the turning effect of forces in situations involving a pivot.

Multiple interacting forces
events in which several forces act enabling pupils to predict outcomes.

4.2 Orientation and elicitation activities

The philosophy underpinning the research was that this science content should be accessed through children's expressions of their understanding of their everyday experiences. Where possible the probes used were introduced by the teachers with practical activities aimed at enabling children to focus on the understandings generated by these experiences.

The initial meeting between participating teachers and researchers concentrated on developing a number of activities which would provide adequate coverage of the science content but, nevertheless, be manageable within the resource and time constraints pertaining. In view of the wide age-range of the children to be involved it was recognised even at this early stage that it would be necessary to tailor the probes for two groupings: one for Reception and KS1 children and one for children at KS2 and KS3. Furthermore, it was agreed that individual teachers would yet more finely tune the probes to suit the children in their classes. Assistance with communicating understanding, for example, would be made available to those with limited ability in this regard. The resulting clustering of probes around activities is outlined below.

Types of force

This activity provided the children with an opportunity to indicate their understanding of the word 'force'. They were asked to write or draw examples of forces, indicate the effect of each force and describe how the force achieves this effect.

Wheels

The starting point for this cluster of probes was the children's experience of riding a bicycle. They were asked to describe what they had to do to start a bicycle moving and then to keep it moving at a constant speed thus demonstrating their understanding of the role of force in changing movement, of friction and air resistance in slowing movement and of balanced forces. For KS2 and KS3 children only, the conventional use of arrows to represent the forces on the bicycle was also explored.

The concepts of inertia and momentum were considered in two further questions about car travel. One asked for explanations of the effect on passengers of sudden braking and the other for predictions of the path of a can dropped out of a window.

Children's ideas about reaction forces were elicited by a question concerning two children on roller skates and for KS2 and KS3 only a further question explored their understanding of acceleration resulting from unbalanced forces acting on a school bus.

Measuring

This cluster contained questions based on activities using objects of different mass, a top-pan balance and a force-meter. Children's ideas about the magnitude of forces and the unit in which force is measured were sought. The opportunity was also taken to gain insights into their ideas about reaction forces using the direct experiences of holding objects on an out-stretched hand and of pushing down on a balance.

Children at KS2 and KS3 were also asked to compare weights in newtons with 'weights' marked on packets of grocery items and hence demonstrate understanding of the difference between mass and weight.

Floating and sinking

The basic activity for this probe involved a tank of water and a plastic bottle so weighted that it floated upright. Explanations in terms of forces were sought. KS2 and KS3 children were also asked to explain the apparent loss in weight of a stone immersed in water and to account for the fact that an object which sank in tap water floated in salt solution. Understandings of balanced forces and upthrust were thus elicited.

Helium-filled balloon

Children investigated the movement of a helium-filled party balloon and were then invited to explain why it moved in such ways. Small masses were then added to the balloon until it ceased to move vertically. Children's explanations in terms of forces for the horizontal changes in movement were recorded. Links with understandings in the floating and sinking activity were sought.

Magnets

Children's ideas about non-contact forces were probed in several contexts. In the magnet activity children were asked to explain an arrangement in which a tethered paper clip was held in mid-air by a magnet clamped to the underside of a chair. They were also questioned as to the reason why two magnets aligned side-by-side on a table repelled each other for a few centimetres only. These activities had the potential for the exploration of understanding of the effect of distance on non-contact forces, unbalanced forces and movement, friction as a stopping force and balanced forces in static situations.

Comb and paper

The effect of a rubbed comb on small pieces of paper was investigated by the children. They were then invited to account for the behaviour of the paper in terms the forces causing the movement.

Tennis ball

For this whole-class activity the teacher threw a tennis ball into the air and caught it as it fell. The children were asked to describe the forces on the ball at various positions during its flight. The responses provided evidence of impetus misconceptions, notions of gravity and of air resistance. The teacher threw the ball into the air for a second time, letting it bounce on the table before catching it. The children then drew their understanding of the effects on the ball of hitting the table. Ideas of deformation and of reaction forces were revealed.

Astronauts

A number of questions were posed in the popular context of space travel. This enabled children to display their understandings of gravity, air resistance and the qualitative link between force, mass and movement. KS2 and KS3 children were also given an opportunity to show their appreciation that, in the absence of frictional forces, no force is required to keep an object moving at a constant speed.

In addition to the above, children in Key Stage 2 and Key Stage 3 classes were given a number of pencil and paper questions that were not initiated by an activity. These questions were used to explore a range of aspects not adequately dealt with previously. These aspects included gravity in various situations, pressure, friction, terminal velocity of falling bodies, the magnitude of the newton and the arrow representation of forces.

4.3   Interviewing

The interviews carried out by the researchers of a sample (see Chapter 2) of the children across the three Key Stages used the responses to the above probes as the basis for discussion. The written or drawn annotated responses were used as a focus in each case for the elicitation of more detailed explanations from individual children. Although teachers discouraged 'don't know' type responses to the initial elicitation, a few children remained reticent. The interviews provided additional opportunities to encourage these children to articulate their understanding. Inevitably for some children the interview was a chance for them to rethink and modifications of their ideas compared to their initial expression emerged. This was not a problem in terms of data collection as it was their current thinking which was recorded for analysis.

4.4   Role of the teachers

The participating teachers were enthusiastic about the activities and elicitation probes and readily accommodated them within their teaching schedules. However, a much more difficult problem was the need to exercise considerable discipline over their professional urge to help the children develop their responses whenever they were perceived to be deficient in some way. Teachers feel a tension when pupils express ideas which are 'incorrect.'

4.5   Limitations on content at intervention

Not all of the aspects of the Forces domain explored at elicitation were carried through to the intervention phase. Some, such as pressure, density, and electrostatic forces, were found to be so far removed from the children's understandings that to pursue them would have been profitless. Others were unfortunately axed by considerations of school resources or of the available time. Even for those for which intervention activities were developed, constraints of classroom and laboratory management and the schemes of work in operation at the time led inevitably to less than complete coverage.

5. INTERVENTION

5.0   Introduction

The term 'intervention' is used to describe all those activities to which pupils are exposed with the aim of helping them to develop their ideas; it is a word which encompasses a range of activities as broad as the imaginations of those who invent or introduce them. It is certainly not the intention to convey the idea of intervention as it is used in the experimental paradigm, i.e. a controlled, standardised procedure. The form of intervention used in this research was imaginative and iterative, responsive to children's ideas. It was exploratory rather than very tightly pre-programmed.

5.1   Planning for intervention

Prior to the intervention, the university-based research team was able to make a preliminary survey of the ideas offered by children during the initial elicitation phase. This interim review, prior to the full data analysis, was based on insights gained during individual pupil interviews, observations of children's responses to classroom based work during the elicitation, preliminary review of children's responses to the concept probes and conversations with teachers during this phase. This survey provided a framework on which the intervention could be built. Outcomes of the survey provided a basis for the second meeting with Project teachers in which plans for intervention were further developed.

Intervention is a phase in which teachers help children transform or develop their understanding. Teachers respond to children's ideas with additional activities, the intention of which is to help children develop their ideas further. To do this teachers have to be aware of:

children's starting points,
directions in which learning might proceed,
a range of teaching and learning strategies,
domain-specific subject knowledge.

At the second research group meeting teachers expressed concerns in each of these four areas. In response, they were provided by the university-based researchers with an intervention booklet which addressed these four elements. The booklets comprised individual planning grids each dedicated to sub-themes or micro-domains. The planning grids described: i) children's conceptual difficulties; ii) a hypothesised description of the likely direction of conceptual development; and iii) an indication of possible intervention strategies. Teachers were encouraged to check the descriptions of children's beliefs against ideas held by children in their own classroom. To assist them in this process, the relevant concept probes used during the elicitation were referenced as a source to which teachers could return for evidence of their pupils' ideas. An important feature of these intervention booklets was that they attempted to make explicit the link between children's ideas and possible intervention activities. The inclusion of additional material to help teachers to develop their own understanding of the domain resulted in a comprehensive intervention booklet.

There was no expectation that teachers should address all the sub-themes identified for possible intervention, or that they should address all the activities within a sub-theme. Indeed, teachers were discouraged from attempting to implement all the possibilities described in the booklet in their entirety. It was important that teachers crafted the intervention to suit the needs and interests of their pupils. Teachers were exhorted to use children's ideas to determine starting points and it was for teachers to judge the point at which to halt the intervention. To assist them in their judgements of appropriate starting points, indications of National Curriculum levels were attached to a description of the anticipated course of conceptual progression which had been outlined in the Booklet. In order to help teachers with the issue of sequencing learning, the five sub-themes were presented in a hypothetical order of intellectual demand. To provide additional support, these five areas were linked to one or more of the three Key Stages. Within these broad indications of Key Stage appropriateness teachers were invited to air their own judgement in the selection of particular sub-themes as foci for the intervention.

5.2   Division of intervention into sub-themes or micro-domains

The brief review of children's ideas provided insights into possible starting points for intervention. The resulting clustering of ideas also formed the basis for structuring the intervention. Five areas in which children's ideas might be most profitably developed were identified. These were:

Forces and their effects on movement;
gravitational force;
friction and air resistance;
reaction forces;
multiple forces, balanced and unbalanced.

5.2.1   Forces and their effects

Firstly, it was anticipated that children would benefit from help in refining their ideas as to what counts as a force, together with an attempt to extend their understanding of the effects of forces. Children across the three Key Stages showed limited understanding of what counts as a force. At the same time they expressed limited understanding of the *effects* of forces. There was little appreciation amongst younger children that forces start things moving. Across Key Stages 1, 2 and 3 there was little awareness that forces are essential to slowing down, or stopping movement. The effects of forces on changing the shape of materials tended not to be appreciated. Children's experience of the quantification of forces was probed at Key Stages 2 and 3. On the whole, these children did not appreciate that forces could be quantified and measured in newtons. They seemed not to understand that changes in the direction of movement are not possible without a force, or, forces being applied. There was no obvious appreciation of the effects of turning forces on movement, or of the concept of pressure.

Intervention focused on what counts as a force and the effects of forces on objects. Key Stage 1 children explored whole body movements to extend their understanding of forces as pushes and pulls and to resolve confusions which had become apparent amongst this age group between a push and a pull. The introduction of sequencing forces according to estimations of magnitude and non-standard measurement of forces was thought to represent a useful preparation for more formal quantification. Key Stage 2 and 3 children also considered what counts as a force but extended this work to carry out some practical measurement of forces using force-meters. These ideas were developed further by considering different sized forces and their effects.

5.2.2   Gravitational force

The second area of development was that of children's understanding of gravitational force. The term 'gravity' was used by some children in Key Stages 2 and 3, but not always appropriately. Their explanations of gravity seemed to be extremely context-sensitive with different instances provoking differing explanations. Children sometimes appeared to hold more than one, often inconsistent, notion of how gravity works, suggesting for example that it can both push and pull objects down. Most had a localised, ground-centred, or Earth-centred view of gravity suggesting that it operated on Earth but was not operating in space, or more specifically not operating on the Moon. Children at Key Stage 2 and Key Stage 3 draw on their everyday experience of weight and frequently distinguished between gravity and weight identifying them as separate phenomena. Few associated gravity with the mass of objects. It was therefore of interest to see whether intervention might help children develop more helpful explanations of the *nature* of gravity which would enable them explain the *effects* of gravity in different contexts.

5.2.3   Friction and air resistance

A third area for development was children's understanding of friction and air resistance. Children demonstrated confused use of both terms. Friction was often associated with objects such as handlebars, pedals, or wheels of bikes and buses. However its precise role in the movement of these objects was less clear. While many children suggested friction was synonymous with 'grip', fewer were aware that it slows down movement, or that it can occur if there is no movement of one object relative to another. Children's understanding of air resistance was often confused by their everyday experience of wind moving. Their use of the term 'wind resistance' seemed to constrain appreciation of the nature of the resistance due to movement through static air. Intervention used empirical investigations of the effects of air resistance and friction to raise children's awareness of the effects of these forces.

5.2.4   Reaction forces

A fourth area of intervention was children's understanding of reaction forces. While children cited numerous forces in different contexts, most focused on separate forces rather than forces in pairs. Intervention through empirical investigations, direct experiences, social challenges and bridging analogies aimed to help children to understand the behaviour and effects of reaction forces.

5.2.5   *Multiple forces*

The fifth cluster of ideas upon which intervention focused was children's understanding of multiple forces and their effects. Children across Key Stages 2 and 3 could cite numerous forces operating in different instances. Most were able to suggest that more than one force could operate on an object at a particular time. However, the effects of these forces were often considered separately. Their effects in combination were rarely appreciated. The concept of balanced forces on stationary objects was not well understood and was often conceptualised as a resolution of a physical struggle, the stronger force overcoming the weaker, rather than in terms of equivalence of magnitude of forces. A similar explanation was often offered to explain the movement of objects. The upward movement of objects might be explained in terms of the *lightness* of the object, or of the upward force *overcoming* the force downward. Movement downward might be attributed to the *heaviness* of an object, lack of *support,* natural *inclination* or the *strength* of the downward force. Rarely were unbalanced forces advanced as explanations for changes in the movement of objects. Intervention in the form of direct experiences, social challenges and empirical investigations attempted to help children appreciate the effect of unbalanced forces on the movement of objects and to develop an awareness of balanced forces acting on objects moving at constant speed.

These sub-themes were sequenced hierarchically so that children at the Lower Key Stages, for example would be expected to consider only the first sub-theme, that of forces and their effects.

5.3   *Evidence of intervention strategies*

There are several sources of evidence as to the intervention strategies adopted in different classrooms. Video-recordings and field notes of visits to classrooms provide a rich and detailed source of data. This evidence also captures something of the situated nature of the intervention strategies and some detail about the ways in which teachers operationalised the intervention suggestions. Important insights into children's emerging understandings are revealed by these data. Insights into the teachers' aims and practices are additionally available from their narrative descriptions of particular instances of intervention. Some teachers indicated how activities were tailored to particular ideas. Children's work reveals evidence of the range of interventions and provides important evidence about children's responses to the learning activities.

Teachers and researchers working on previous enquiries under the umbrella of the SPACE project as well as other collaborative research at CRIPSAT, had developed a number of strategies for intervention. The research team was also aware of recent developments in understanding the nature of cognitive development. The influence of social processes on cognition is well documented by proponents of the socio-cultural views of Vygotsky (1978), and in the situated cognition literature (Chaiklin and Lave, 1993; Lave and Wenger, 1995). More recent interest in cognitive-developmental psychology, has focused on the representational form in which concepts might be constructed, stored and accessed (Karmiloff-Smith, 1992). Karmiloff-Smith hypothesises that cognitive development proceeds through a process of representational re-description. It is suggested by Karmiloff-Smith that representations are stored in different modes which are more or less accessible to the system. Through the organising modality of language. This theory is of importance to the theoretcal understanding of project activities and learning outcomes for several reasons. Firstly, the elicitation phase provided an opportunity for children to make explicit some of their representations. Secondly, the task for the teacher might be described as one in which he or she helps children make their representations explicit prior to constructing links between these different representations. Thirdly, there are important implications for the different sources of knowledge which may be accessed via the classroom and the modes in which knowledge is shared and acquired. It is suggested that teaching and learning strategies which encourage children to re-describe their understanding in different modes might enhance the possibility of conceptual progression. Teachers were therefore encouraged to adopt a diversity of approaches using a variety of modes through which knowledge might be made explicit by the learner. While the project team was interested in particularly effective teaching and learning strategies, it did not assume that 'one-off', closely targeted interventions which could assure conceptual progression would be the outcome. Redundancy in teaching strategies was anticipated and indeed will be regarded as functional once it is accepted that knowledge can be represented in different modes within an individual's cognitive processing system.

Intervention drew on a range of modes of activity which were made explicit to the project teachers. The provision of opportunities for social challenges in the form of small group or whole class discussion was suggested. An opportunity to make ideas explicit and discuss the evidence for those ideas was considered likely to provoke some development in understanding. Empirical investigations were encouraged as a way of children gathering and evaluating the evidence for their ideas. Secondary sources of evidence, for example from videos, was also provided where necessary. Bridging analogies were used to try to help make accessible those visually imperceptible aspects of forces. Creative writing was used as a means of asking children to consider improbable events such as the absence of friction. These improbable events encouraged children to move from the concrete to the abstract. Direct experiences together with some evaluation of the evidence and ideas gained from such experiences were used as a means of helping children to develop their ideas.

The interactive nature of the intervention was important, since it was through teacher-pupil interactions that teachers could monitor children's developing understanding in terms of conceptual progression and determine the shape of subsequent learning activities. A second important feature of intervention was children's active involvement in the accumulation and evaluation of evidence. While teachers ensured that different sources of knowledge were made accessible, it was for the children to decide what counted as evidence in support of (or as a challenge to) their beliefs. The teachers' role was to introduce children to sources of knowledge and help them make sense of that evidence against the background of their own ideas, or the pupil representations generated in the elicitation phase.

In the following sections of this chapter each sub-theme targeted for development in the intervention phase is presented in turn and the nature of the intervention undertaken in the area reported. Some of the sub-theme discussions review evidence of interventions in all three Key Stages; others address interventions explored by just one or two of the Key

Stages. The intention is to give some indication of the kinds of activity which took place together with a sense of some of the outcomes of those activities. It does not reflect the work of any one class but rather, through describing some of the quality of what happened in different classes, presents a general picture of intervention. For any individual teacher it was possible to tackle only a fraction of the spectrum of activities described over the following pages.

5.4 Intervening to help develop ideas about forces and their effects on movement

Many children tended not to conceptualise 'forces' as actions such as pushes and pulls so much as to associate them with objects. Additionally, young children revealed confused understanding of pushes and pulls. This sub-theme was judged to be of primary importance to children's developing understanding of the domain and was therefore a feature of intervention for children in all three Key Stages. Teachers approached this sub-theme in different ways, developing and selecting activities according to the needs of their particular class. Some aspects were briefly addressed by classes of older children while the sub-theme dominated the intervention experience of children at Key Stage 1.

All of the teachers working with children in Reception and Key Stage 1 appreciated the value of children firstly considering pushes and pulls in relation to their own bodies and in the context of everyday familiar activities. An important aspect of the teaching and learning activities at Key Stage 1 was distinguishing between a push and a pull. The example below illustrates how one teacher of Reception and Year 1 children worked with a small group of children, exploring their ideas about pushes and pulls involved in getting dressed.

Tch   *How did you put on your hat?*
Ch 1   *I pulled on the hat and then pushed it down. To take it off I need to pull it.*
Ch 2   *I pulled off my cardigan.*
Tch   *How are you taking off your shoes Zoe?*
Ch 3   *I pushed first, then pulled them off.*

The teacher encouraged children to demonstrate the pushes and pulls which they had described. A Year 2 teacher who used the same activity focussed the discussions on the words children used to describe their actions. Discussions had elicited 'shove', 'push', 'pull', 'slide'. To help children link these actions to pushes and pulls and begin to label their understanding the teacher asked

Tch   *Can you think of two special words we have been using?*
Ch   *Pushing and pulling.*

Children demonstrated the various actions and group discussions clarified whether the examples were indeed pushes or pulls. They made drawings of their ideas listing actions as pushes and pulls. (See Figure 5.1).

As the teacher used the same activity with different groups of children, she noticed some variation in the ways in which different children approached the task. Some children experienced difficulty generating examples of pushes and pulls. Others readily generated pushes and pulls and observed that some actions involved both pushes *and* pulls. For these children, subdividing actions into pushes and pulls was insufficient; they recognised the need for a third category in which actions involving pushes *and* pulls might be placed. The discussion below shows how other children; with teacher support, gradually came to a realisation of the need for a third group as they were completing their individual drawings.

Figure 5.1

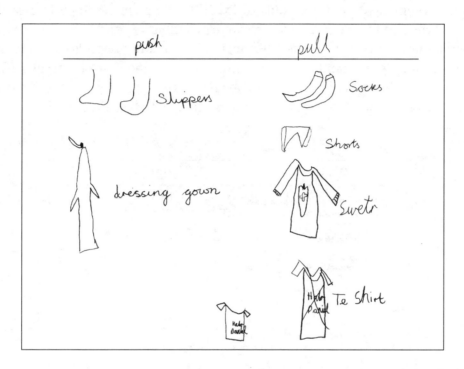

Y2 B M*

Tch	*Did you find that the things you drew fit nicely on one side or the other?*
Ch 1	*Boots, you push your feet in and pull your zip up.*
Tch	*What do you do with those things that are a push and a pull?*
Ch 1	*You could put them in both sides.*
Ch 2	*You could put them half in each side.*

The link between pushes and pulls and personal actions was generalised through direct experience of pushes and pulls in other familiar contexts such as playing and PE. Children discussed examples and then made drawings of their ideas. A focus for their activities was the encouragement of discrimination between pushes and pulls. The teacher noted that children seemed to experience more difficulty generating pulls than pushes. A Year 3 and 4 teacher described how class discussion helped children distinguish between pushes and pulls. Children had been asked to show, on a work sheet, whether different actions were pushes or pulls. Discussion revealed points of difference in children's reasoning and according to the teacher, promoted the possibility of changes in children's ideas.

* These codes identify Year group, Gender and overall achievement, High, Medium and Low.

We then discussed some of the ones where children had different answers. This was useful as children seemed to be consolidating their ideas as they talked about them, or also to change their ideas as they thought about them and as other children were explaining why they had put either a push or a pull. The children came round to deciding that cycling was both [a push and a pull] after various children tried to explain-as some had put 'pull' and others 'push'.

Year 3/4 teacher

It was important to the teachers that children's understanding of forces should be built on the familiar. Many commented upon the value of experiencing pushes and pulls. However, many also recognised the need to extend children's understanding beyond human actions. The work in several classrooms was extended so that children could begin to consider non-human pushes and pulls. While the younger children experienced some initial difficulty generating pushes and pulls exerted by non-human agents, the lists of pushes and pulls generated by a Year five child shows several non-human examples. (See Figure 5.2).

Figure 5.2

Push	Pull
Swimmer - forwards, friction and gravity and muscle power.	Swimmer - forwards, friction and gravity and muscle power.
water - anywhere, gravity	water - anywhere, gravity
Boat - forwards, gravity, friction.	
Rocket - forwards, upwards, gravity friction.	
Car - forwards, backwards, friction gravity	
Person walking - forwards, backwards sidewards, friction, gravity.	

Y5 G H

Teachers indicated that, at Key Stage 1, the progression from human pushes and pulls to non-human pushes and pulls was a novel approach which they found useful. For children at Key Stage 1 a particular difficulty was the generation of examples of pulls and confusion between pushes and pulls. Teachers in other Key Stages felt the initial focus on the nature of pushes and pulls and their effects was necessary to support later learning in the domain. Helping children appreciate the *effects* of pushes and pulls proceeded in three directions. One way was to consider the effects of pushes and pulls on *movement*. The second was to consider the effects on the *shape* of an object. A third was to consider variation in the *size* of forces and their effects. These areas of understanding were taken up by teachers at Key Stages 1, 2 and 3. The first two areas provided a particular focus of intervention at Key Stage 1 and Lower Key Stage 2 and were used as an important introduction for further work at Key Stages 2 and 3.

5.4.1   Helping children understand the effects of forces on movement of objects

A teacher of Year 1 and Reception children noted that children's work on pushes and pulls revealed that some children did not recognise that pushes and pulls start *movement*. She encouraged groups of children to find ways of making a toy car move (see Figure 5.3). These direct experiences were quickly developed to consider ways of making the car move faster and ways of making it stop. This activity aimed to help children make a direct link between pushes and pulls and changes in the *movement* of an object.

Figure 5.3

In a Year 2 class the teacher provided practical experiences of different ways to start objects moving to help them discriminate between a push and a pull. Children explored moving objects on land and water, by pushing using hands or a stick, and by blowing, using a straw, or a battery-powered hairdryer. The teacher's questions aimed to help children to appreciate that pushes and pulls can operate on the same object, both resulting in the same outcome in terms of movement. Also, some objects could exert a pull but not a push (e.g. the pull of a string) while others could push but not pull (e.g. the jet of air from the hairdryer).

Was it a push or a pull?
Can you pull the boat using the hairdryer?
Can you push the boat with the string?

At the end of the activities the teacher noted that while children appreciated that pushes and pulls start movement, some children continued to have difficulty in identifying a force as a pull when the action is away from the child.

Another Year 2 teacher used the movement of balls to encourage children to consider the range of ways pushes and pulls affect movement. The children noted pushes and pulls could make the balls move downwards, upwards and roll forwards and backwards. Annotated notes on drawings indicate the scope within individual discussions for probing how children distinguish between a push and a pull. The Year 2 child's drawing (see Figure 5.4) illustrates the different ways she noticed pushes and pulls affecting the ball. She observed how pushes change the movement and shape of the ball. The opportunity for individual discussion with the teacher also reveals a surprising ability to link the ball's bounce on the floor to the floor pushing back.

Children suggested that pushes and pulls resulted in objects moving in different directions, but they had some difficulty within the intervention appreciating that an object would move in one direction while one person was pulling and another pushing. Direct experiences of moving large pieces of classroom furniture helped children understand that a push and pull could result in movement in the same direction. (See Figure 5.5).

Figure 5.4

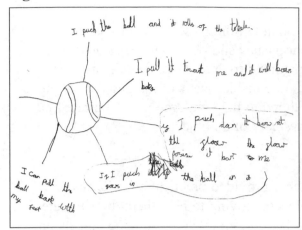

I push the ball and it rolls off the table.
I pull it towards me and it will bounce back.
If I push down at the floor, the floor pushes it back to me.
If I push the ball in it goes up.
I can pull the ball back with my feet.

Y2 G M

Figure 5.5

Y2 G M

Key Stage 2 teachers believed that an exploration of the effects of pushes and pulls on the movement of objects was important preparation for later work on forces. In one Year 3 and 4 class, children were encouraged to think not just about pushes starting movement but ways of *stopping* movement. This teacher encouraged children to use arrows to indicate the direction of pushes and pulls, to emphasise the direction in which forces were acting. (See Figure 5.6).

Figure 5.6

Y4 B M

5.4.2 Helping children to develop their ideas about how forces can change the shape of objects

One way of helping children to appreciate that pushes and pulls can change the shape of objects was to provide opportunities for them to explore objects made of different materials. A Reception teacher provided groups of children with a range of materials including modelling clay, sponge, polystyrene and rubber bands (see Figure 5.7). Children were eager to handle the materials and each in turn described what happened when they gave one of the materials a pull, or a push. The focus within work for this class was the oral articulation of ideas and the development of vocabulary with which children could express their observations. Some of the children struggled to express their observations and could not easily produce words like 'squash' and 'stretch'.

Figure 5.7

One Year 2 teacher incorporated this idea into intervention, aiming to help children understand the wider range of effects of pushes and pulls. She gave groups of children a single object and asked them to draw or write what effects pushes and pulls had on the object. The range of objects across the class included springs, plasticine, sponges and elastic bands. The class discussion brought children's experiences of the different objects together. Ideas of the effects of pushes and pulls on different objects were reviewed and common effects of pushes and pulls were added to a class list.

A Year 3 and 4 teacher encouraged children to explore the effects of pushes and pulls on the shape of materials. (See Figure 5.8).

Figure 5.8

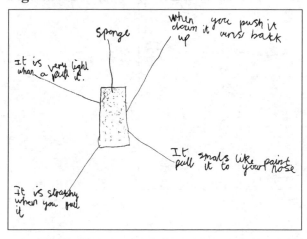

Sponge
When you push it down it comes back up.
It smells like paint. Pull it to your nose.
It is stretchy when you pull it.
It is very light when I pull it.

Y3 G L

Discrimination between pushes and pulls was thought to be important by most teachers at Key Stages 1 and 2. They provided initial opportunities for children to generate a range of pushes and pulls and classify these into two groups. The difficulty experienced at Key Stage 1 of generating pulls and of linking pushes and pulls to the movement of objects was not so apparent at Key Stage 2. Nevertheless, they were considered to be a useful way of linking children's direct experience of pushes and pulls and their effects to more abstract consideration of forces in subsequent lessons.

5.4.3   *Helping children to understand quantification of forces*

The idea that forces can be quantified was introduced to children in different ways across the three Key Stages as children engaged in activities related to a gradual appreciation of measurement of forces. Different starting points and stopping points could be identified and a tentative sequence of activities tailored to children's developing understanding can be mapped out. Interventions in this area will be presented in Key Stage order to demonstrate areas of difference and of overlap.

In some Key Stage 1 classrooms it was possible to extend children's initial consideration of pushes and pulls to help them appreciate that the *size* of pushes and pulls can vary, together with an appreciation that different sized forces will have different effects. It was believed that intervention in this area could help lay the foundations for later appreciation of the quantification of forces.

In a Year 1 class the teacher asked children to think of large and small pushes and pulls. A class list of their ideas was constructed following class discussion. They clarified what counted as a large and small push/pull. Later children investigated the effects of different sized pushes and pulls on different objects. The drawing reproduced as figure 5.9 shows how a Year 1 child identified and represented the effects of different sized forces on an elastic band and the movement of a car.

Figure 5.9

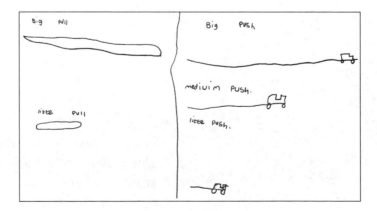

Y1 G H

The class teacher helped one group to quantify the effects of different sized pushes using non-standard measures. These investigations helped children begin to understand the relationship between the size of the force and its effect.

A different way of exploring different sized pushes is to consider how much force is needed to start different objects moving. In a Year 2 class, a few children were given three objects of different masses and asked to sequence the objects according to the size of the push which would be needed to make them move. The children's drawing (see Figure 5.10) together with their expression of ideas shows that they appreciated the relationship between the heaviness of an object and the size of the push needed to make it move.

Figure 5.10

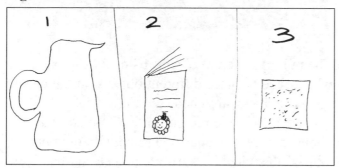

The jug is first it needs a big push.
The book needs a middle sized push.
The sponge needs a little push.

Y2 G M

Many Key Stage 2 children were unfamiliar with newton-meters despite them being available for use in many classrooms. In order to lead children towards an appreciation that forces could be quantified, some Key Stage 2 teachers encouraged them to make initial qualitative judgements about the relative sizes of forces. In one Year 6 classroom, children were encouraged to make their own lists of forces together with estimates of their size. Figure 5.11 below illustrates one such example in which the magnitude of each force is labelled on a three-point descriptive scale: 'small;', 'medium' or 'big.' This reflects the demand of the elicitation probe used in the research and is clearly capable of being extended to more elaborate series, moving to quantified estimates in newtons.

Figure 5.11

elastic band	pushing force	medium force
elastic thickband	pulling force	big force
thin elastic band	pulling force	small force
plastic skipping rope	pulling force	small force
rope skipping rope	pulling force	big force
1	2	3.

Y6 G M

48

Figure 5.12

Figure 5.13

In another classroom, children investigated the pull of different objects using elastic bands attached to hooks, (see Figure 5.12). They were asked to place different objects on the elastic bands and to measure the length to which the elastic band stretched (see Figure 5.13). Establishing a link between the length of the elastic band and the pull of each object was no easy task. The teacher questioned children about their results and tried to help them appreciate the relationship between the pull of the object illustrated by the stretched elastic band and the weight of each object. As well as offering experience of a perceptible outcome of the effects of different weights, children would also be helped to consider the parallels between the elastic band and the spring in the force-meter.

A teacher of a Year 3 class planned a sequence of activities which supported the development of children's ideas from an awareness of pushes and pulls, to an appreciation that forces could be measured. Firstly, children rank-ordered a selection of pushes and pulls. They then made a push/pull meter which enabled them to take non-standard measures of those same pushes and pulls. The teacher's account indicated that children used this device in a variety of situations, just one of which was weighing.

Encouraging children to make and use a push/pull meter was valuable in two respects. Firstly, it provided the possibility of measuring pushes and pulls using non-standard measures. Secondly, it offered the possibility of measuring pushes *and* pulls in contrast to conventional spring-balance type newton-meters which measure only *pulls*. Using a push/pull meter provides a bridge between the qualitative terms such as big/little push used by younger children to discriminate sizes of pushes and pulls and the more formal measurements in newtons offered by a newton-meter. Non-standard measurement helps children understand what they are measuring, how the measuring device works and the need for standard quantification of measurements. It is anticipated that this sequence enhances the possibility of children being more receptive to the introduction of newton-meters. The teacher reported that, when they children were introduced to the newton-meter, they were in a position to understand the function of the device and the need for accurate standard measurements in their investigations of forces. Records of children's investigations show their appreciation that the newton-meter was measuring the pull of an object. (See Figure 5.14).

In a Year 6 class, children made the push/pull meter according to the suggestions of the researchers and used it to investigate how much push was needed to make different objects start moving. The children developed this investigation themselves as a result of making the device. Children's ideas could be further developed by using their new-found awareness of the uses of standard measurement of pushes and pulls to calibrate their own device against a standard one. (Alternatively, they could be helped by the provision of the information that a force meter pulls upwards with a force of 1N when a mass of 100 g is suspended from it. Activities of this kind can help scaffold the transition to measurement using newton-meters).

Figure 5.14

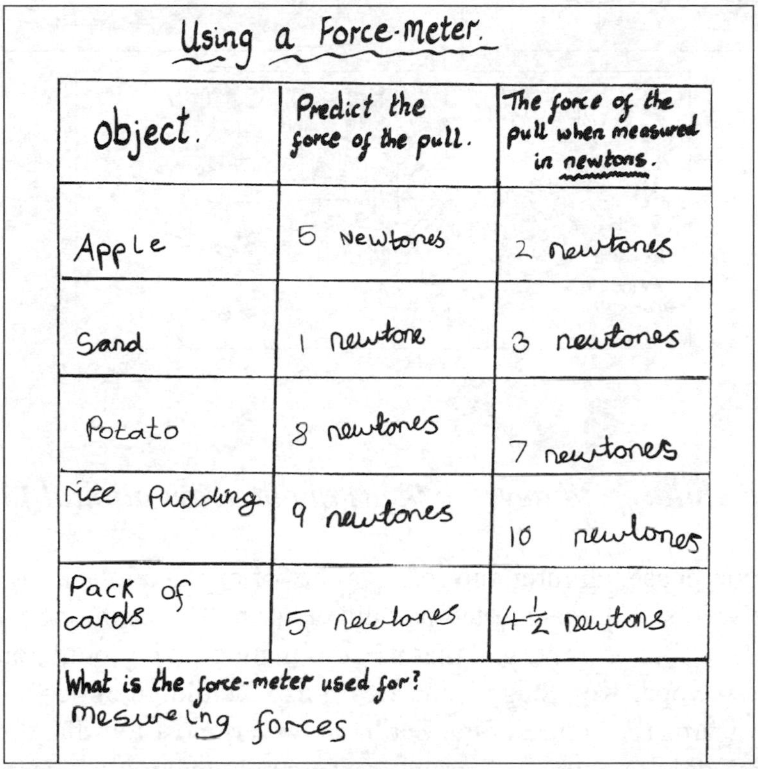

Using a Force-meter.

object.	Predict the force of the pull.	The force of the pull when measured in newtons.
Apple	5 Newtones	2 newtones
Sand	1 newtone	3 newtones
Potato	8 newtones	7 newtones
rice Pudding	9 newtones	10 newtones
Pack of cards	5 newtones	4½ newtons

What is the force-meter used for?
mesureing forces

Y3 G H

Introducing children to formal systems for *measuring* forces is one aspect of conceptual development. Another is to introduce children to the culturally accepted symbols for *representing* the size and direction of forces. Many children used arrows to show what they believed to be the *location* of a force rather than to indicate its *direction* or *magnitude*. In a Year 3 class, the teacher asked children to show what they believed to be the direction of forces using arrows. The drawings show children's ability to represent direction of their own pushes and pulls (see Figure 5.15).

The teacher believed that consideration of the *direction* of forces was at the limits of the Year 3 children's understanding and that representation of the *magnitude* could wait until children had much greater experience of measuring forces.

Figure 5.15

Y3 G M

5.5 *Helping children to develop ideas about gravitational forces*

During the elicitation phase, children offered a variety of explanations as to how the Earth's gravitational force works. They suggested it could either pull or push or both pull *and* push . Most understood the effect of gravity in that it keeps them on the ground, or stops them floating away. A few Upper Key Stage 2 and Key Stage 3 children understood that the direction of gravity is towards the centre of the Earth. However, most felt that there was no gravity in space and most did not appreciate gravity as something acting *between* two objects. Children's explanations of how gravity works revealed aspects of their understanding which needed further support. Some of the children believed gravity operated on land but not in water and ideas that gravity was linked to air were also revealed. One of the ways of helping children to develop their ideas about gravity was considered to be to help them appreciate gravity as a force that is directional. A Year 3 and 4 teacher asked children to draw pictures and show the direction of gravity using arrows. To help children appreciate that the force due to the Earth's gravity is towards the centre of the Earth, children were given globes and stuck figures at different points on the Earth's surface. The teacher felt that the discussion around these models was useful in helping children towards more Earth-centred notions of gravity appropriate to the age group.

We used a globe and stuck model figures on it in different places and I asked, 'What is it that makes people stay on the Earth in different parts of the world?' They said that it was gravity, coming from the centre of the Earth.

Year 3/4 Teacher

To check their appreciation that the Earth's gravitational attraction operates towards the centre of the Earth, she gave children two pictures, one of a tank of water and one of the Earth. Children were encouraged to draw arrows on both pictures to show the direction of gravity.

The child's drawings reproduced in Figure 5.16 demonstrate his appreciation that gravity acts towards the centre of the Earth.

Figure 5.16

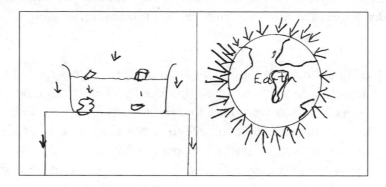

Y4 B H

In a Year 5 and 6 class, a teacher used the analogy of magnetic attraction to suggest that gravity was a force of attraction pulling down. The teacher explained:

This level of teaching input and teacher directed activity was necessary as a foundation for the next experiences.

Year 5/6 teacher

The teacher's notes indicate a later sense of dissatisfaction with some aspects of this direct teaching. As with all analogies and models, some attributes of the analogy map across usefully to support science understanding. However, not all attributes of the model are equally useful. On balance, this teacher came to the conclusion that the analogy between the gravitational attraction of the Earth and magnetic attraction was not entirely helpful. Some children, she felt, were confused by the parallel.

I did use the idea of magnets to talk about attraction and this is the idea that has stuck in their minds. This is something I must remember as I feel it has confused the children.

Year 5/6 teacher

There are some aspects of the analogy which the child was able to cling to and which are useful, children might need further discussion of some aspects, such as whether gravity is an artefact like a magnet, or a force between two objects. On balance, the analogy linked some familiar experiences to the newly presented knowledge. In addition, both magnetism and gravity are examples of non-contact forces. While the analogy was limited in many respects it led to some developments in understanding, such as gravity being conceptualised as a pull *towards* the Earth.

Teaching and learning activities adopted in these Key Stage 2 classrooms aimed to help children to understand the direction in which the Earth's gravitational force acts, perhaps helping them move from a 'ground-centred' to an 'Earth-centred' view. The idea of gravity as the force of attraction between two masses tended not to be addressed by teachers at Key Stage 2.

A preliminary stage might be to help children distinguish between mass and weight. A teacher of a Year 5 and 6 held a class discussion to elicit ideas about the nature of weight and mass in the course of which, children's everyday understandings emerged. The teacher introduced children directly to the idea that mass is 'how much stuff something is made of'. Children's everyday sense of weight was used as an introduction to weight as a force downwards.

Interventions addressing the mass/weight distinction at Key Stage 3 tended to be in the form of class discussions in which new ideas were introduced and existing ideas put to public scrutiny. In some lessons, teachers were able to use direct experiences, video evidence or children's investigations to provide the evidence to fuel the debates. In one Year 7 class, children were encouraged to 'weigh' themselves using kilograms and newtons. Their task was then to consider how the two measurements might change if they took the same measurements on the Moon. The teacher hoped that students might be helped to understand that they were measuring different properties. Children worked individually on this task, followed by a whole class discussion which provided an opportunity to share results and interpret data.

5.6 *Helping children to develop their ideas about friction and air resistance*

During the elicitation phase Key Stage 2 and 3 children showed awareness of friction when they mentioned it as a force. They tended, however, to associate it with particular objects, almost as though it were a property of those objects. There was some confusion about whether friction was operating in different instances and how friction worked. The intervention aimed to follow up these ideas through direct experiences of instances of friction. In one Year 5 and 6 class, the teacher provided some direct experiences of friction for children to experience directly and tangibly and reflect upon. One of the instances was a rubber pad designed to help the elderly or those with problems in gripping to open the lids on jars. Children were asked to explore whether there was a difference in opening the jar with wet hands and with the pad. While all the children noted the ease with which the jar opened using the pad compared to wet hands, they differed in the extent to which they understood friction as playing a role. Children recorded some of their ideas in writing (see Figure 5.17).

Figure 5.17

Example 1

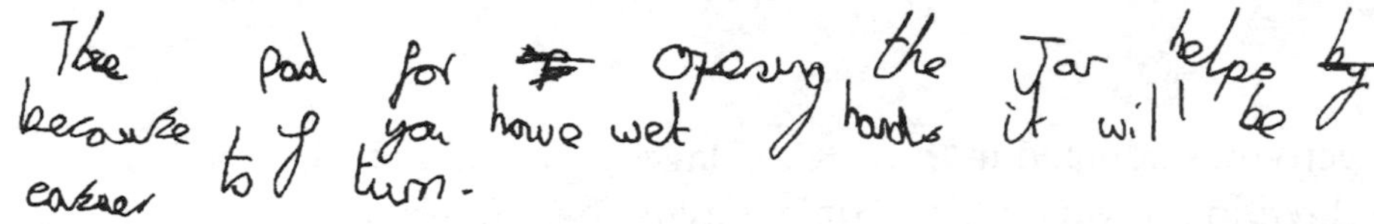

The pad for opening the jar helps because if you have wet hands it will be easier to turn.

Y5 G L

Example 2

The rubber pad helps you if you wet hands because I think rubber provides more grip

Y6 B M

Example 3

The pad decreces the griction on the jar lid making it eysier to open and close tightly. Rubber is the best type of material to use because it is glexable and bendy.

Y6 G M

Example 4

The pad for opening the lids of jars work by the rubber and the lid rubbing together and causing friction and therefore making it easier to open the jar. Rubber is a good material for bending, it is flexible for small jar & big jar lids.

Y6 G H

The children's written responses records the different ways in which they interprctcd the experience and their explanation as to why they thought that the pad makes the lid easier to turn. Many suggested that the rubber provides more 'grip' an expression which might be interpreted as an approximation to the idea that friction is operating. If this is so, it might be helpful to provide the scientific label as an alternative to, or substitute for, the vernacular expression of the idea. Indeed one teacher commented upon the need to use both 'grip' and 'friction' during discussions in order to help children label their understanding as friction.

I had to maybe reinforce the idea that more friction meant more grip, just by giving examples and using the word 'friction' as well as 'grip' all the way through.

Year 5/6 teacher

The child who offered Example 1 in Figure 5.17 does not use the word 'grip', though Example 2 does. The latter might be an example of a response from a child who is ready for the correct scientific vocabulary. Example 3 is more difficult to interpret because the respondent refers to a *decrease* in friction, and there is indeed friction between the lid and the glass of the jar. Such responses can provide the stimulus for fruitful class discussion. Example 4 is the only one of those reproduced here which introduces the idea of friction as a force operating *between* two surfaces.

54

The teacher helped children generalise their ideas by considering a second context, that of the brakes on a bicycle. Pupils had an opportunity to experience the cycle brakes in operation before making explicit their ideas in response to the instruction, 'Describe how a brake block on a bicycle manages to stop the wheel turning.' (See Figure 5.18).

Figure 5.18

> The rubber block presses against the rubber tyre and causes the tyre to slow down and finaly stop.

Y6 G M

> The brake block also made of rubber has little zig zag lines which when slamed down on the wheel which increases friction and pulls the wheel to a halt

Y6 B M

The first piece of writing describes observations of the effect of the brake block on the movement of the wheel. These observations are important since they could lead to a later appreciation that slowing down and stopping the movement of an object requires a force and that in this instance the force is termed 'friction'. The second reports that the indentations in the brake block increase friction and slow down movement. It is interesting that friction is described as a pull. Further work might focus on this belief and help the child begin to appreciate friction as a force acting in the direction opposite to the movement of the wheel. While it is useful to explore instances of the same concept in different contexts, the cycle context presents some counter-intuitive complexity because the friction force between the road and the moving wheels is in the same direction as the movement of the bicycle. It might be that other instances which are more closely linked to intermediate understandings might be introduced prior to examination of frictional forces on a wheeled vehicle.

Following a suggestion offered by the research team the teacher encouraged children to reflect on imaginary events in order to help them develop their ideas about friction. As a creative writing exercise, she asked them to write about a world with no friction. They were encouraged to consider the effects this would have. It was anticipated that, in order to consider a world without friction, children would first need to reflect upon their understanding of how friction works and the effects of friction in the real world. The creative writing provided a supportive context for this imaginative leap. Often, children's first thoughts were of examples associated with their direct classroom experiences. This demonstrated that they were linking their ideas from these earlier experiences with the demands of the new task. Many children generated instances linked to their experience of the effects of the brake on a moving cycle, suggesting that in a frictionless world, cars and other vehicles would not stop. Interestingly, there is no evidence to suggest children believed friction was needed to start wheels turning. Others thought of events which would affect their own physical actions. Hands would slide rather than rub together, movement would be uncontrolled and chaotic. The teacher used a class discussion to share the range of instances about the effects of having no friction. These creative writings were effective in encouraging children to recognise instances in which friction *does* operate and to consider its effects. The class discussion drew some of these ideas together and a composite story was constructed. This practice allowed children to make explicit their ideas and reflect on the ideas of others. Frictionless events and their effects were considered, inconsistencies in descriptions of effects were exposed and reformulated within the discussion.

One particular way in which children were helped to understand friction was by collecting evidence from investigations of their ideas. Several Key Stage 2 teachers encouraged children to explore the movement of objects across different surfaces or the movement of different objects across the same surface. The arrangement illustrated in the example drawn from a Year 3 class (reproduced as Figure 5.19 below) allowed children to gather data which they then used to help them in drawing conclusions.

Figure 5.19

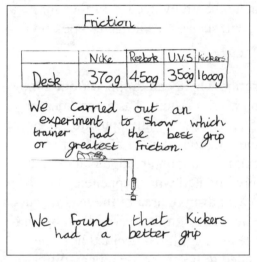

Figure 5.20

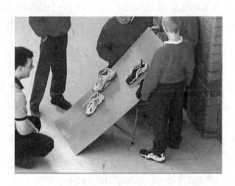

Y3 B M

A Year 6 teacher encouraged the development of children's use of the term 'grip' through the means of an investigation to see which trainer had the best grip. (See Figure 5.20). Children compared the movement of trainers down a smooth slope. Some of the children explained that particular trainers had more grip and therefore more friction. They seemed to believe friction was an intrinsic property of the rough sole. These kind of activities could further develop children's understanding by inviting them to consider the outcomes of their investigations in terms of forces opposing movement.

In another Year 6 class, the teacher asked children to investigate the movement of wooden blocks across different surfaces. They were encouraged to measure the force required to pull the blocks along. Some of the children linked their results to the texture of the different surfaces. A few, like the child below, were able to link their results to friction caused by movement between two surfaces. (See Figure 5.21).

Figure 5.21

Conclusions

I found that the heaviest objects didn't always have the
largest amount of friction newtons. For example, the big
block took less force to move across the desk than the
medium block did. Friction works more on rough surfaces
and it works in the opposite direction to the moving object

Y6 G H

Investigations such as these provided children with evidence about the variations in frictional force. Some of their causal explanations showed an appreciation of forces shift away from a notion of grip as an intrinsic property of only one of the surfaces. Once children appreciated that friction occurs between surfaces moving against each other, they needed further support to understand those instances of friction without movement, such as a stationary book on a slope.

Within the elicitation phase, children were asked to describe why a can dropped from a moving car falls in a particular manner. Many children used 'wind resistance' as a core idea in their explanations of the movement of the can. A Year 5 and 6 teacher attempted to extend children's understanding of stopping forces by asking them to consider the effects of air resistance on objects moving through air. She wanted children to gain an appreciation that movement might be slowed down by the air. Children investigated what happened when a large piece of cardboard was dropped, first on its edge (with a small area of the leading surface) then with its largest surface area falling first. The investigation was carried out by children using the organisational structure of a whole class lesson. Children decided how they would measure the height from which the cardboard needed to be dropped and any difference in the time taken to reach the ground. Other concerns of fair testing, such as controlling the height and point of release and number of trials needed were discussed and selected by the children. Children timed how long the card took to fall, in each trial. Recordings were made on the blackboard and shared with the class. A discussion was organised to reflect on the evidence and for the generation of explanations as to why the card fell at different speeds.

In another Year 5 and 6 class, the teacher followed the investigation of the falling card with other opportunities to investigate air resistance. Two pieces of paper (both pieces were of the same mass, one crumpled into a ball and one open) were released and allowed to fall through the air. The teacher dropped both pieces of paper simultaneously from a height of two metres. Children observed which landed first. All the children correctly observed that the crumpled paper landed first but, they found it difficult to generate an explanation for this observation. A few suggested that the smaller paper weighed more. Interestingly, a few suggested that the crumpled paper was smaller and therefore lighter. These responses suggest a failure to conserve the mass of the paper. The second view is contrary to widespread beliefs expressed in the elicitation that heavy things fall faster. The teacher's description shows her attempts to provide further evidence to challenge some of the children's ideas.

The children seemed to grasp the reason why the card on its side fell more quickly. But with the paper many of the group said it was because the scrunched paper was heavier. I unscrewed the ball and showed them it was the same piece of paper but they still said that when you scrunch it, it still gets a bit heavier. We talked about the hammer and the feather and I said there was no air on the Moon and which did they think would hit the ground first. James said, 'At the same time.' The others said, 'The hammer because it is heavier.'

Year 5/6 teacher

The teacher helped the children make links between the evidence from their investigations and the video evidence of Neil Armstrong's hammer and feather experiment on the Moon, in which both land at the same time. They were urged to consider the link between the absence of air on the Moon and the role of air in slowing down objects moving on Earth. The teacher's description shows that these ideas are accessible to some children. It also shows the powerful influence of their intermediate understandings on children's interpretation of evidence. For these children, associations between the differing contexts were difficult to make and further evidence of the movement of objects might be needed before notions of air resistance become accessible. Until children are able to understand the role of gravity on falling objects, movement downwards might not be the most effective context in which to consider the effects of air resistance. The ideas and practices of this teacher exemplify how a diversity of approaches might be used to help children develop their ideas within a concept area. The different approaches used different modes of activity, with the teacher helping children to construct and access representations through different cognitive modes. Her constant comparison of different instances illustrates one way in which children might be helped to link events and redescribe their representations.

5.7   Helping children develop their ideas about reaction forces

During the elicitation, reaction forces were rarely mentioned by any children. Intervention in this area tended to be addressed by teachers at the Upper end of Key Stage 2 and at Key Stage 3. However, there were some instances of teachers in Years 3 and 4 attempting some introductory work in this area. In general, teachers tended to use whole body experiences to help children understand that the object being pushed pushes back on the thing doing the pushing. It was believed that awareness might be increased if children, when they pushed an object, felt, or gained the sensation of, the object pushing back on them. In a Year 3 and 4, class children were encouraged to pay attention to the feeling of pushing upwards on a book held at arm's length in order to explore the complementary sense of the book pushing downwards on their hand. The teacher felt that this was one way on making reaction forces accessible to young children.

They seemed to be happy with the idea that the book was pushing down and went around the room looking for other objects that might be pushing or pulling.

Year 3/4 teacher

58

Figure 5.22

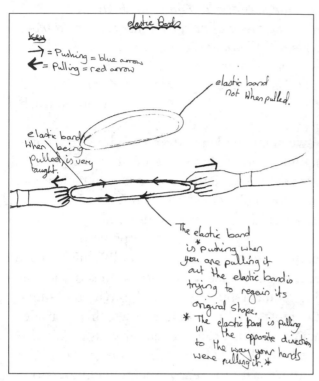

Some teachers asked children to pull an elastic band and consider whether they could feel the elastic band pulling back. A Year 6 teacher asked children to contrast this with what would happen if they moved their hands away from each other without the elastic band. Many of the children recognised that the band was also pulling their hands back together. Children were encouraged to draw their ideas and share these with the rest of the class. Interesting differences emerged in children's beliefs about whether the elastic band pushed or pulled. When children drew their ideas on the blackboard, some showed arrows in the correct direction but were confused in their descriptions of pushes and pulls. Others confused both arrows and descriptions. The class was asked to show their commitment to one of a selection of five drawings and explanations made by their fellow class members. In this way, the teacher helped the children to reflect on their own ideas. Asking children to choose a drawing helped the teacher monitor the changes in ideas following the discussion. The drawing (see Figure 5.22) shows one child's understanding following this sharing of ideas.

Y6 G H

Figure 5.23

Other practical experiences were explored with the aim of helping children to appreciate reaction forces. In groups, children experienced a sequence of three activities. One at a time children sat in a large trolley or wheeled box and pressed their feet or hands on an outside wall (see Figure 5.23). They observed what happened to the movement of the box and tried to explain this event in terms of forces. Some children focused on forces such as friction while others showed that they had at least an intuitive understanding of the wall pushing back.

Figure 5.24

A second activity was for children to explore a range of springs. (See Figure 5.24). These springs varied in size but included some large mattress springs. Children squeezed the springs and considered what forces were acting. They focused on the spring both in a horizontal position (pushing the spring together with both hands) and vertically, (on a table, and pressed down with the hands).

The teacher's questions then probed whether they could feel anything pushing back up. The drawing below (see Figure 5.25) shows how Claire (Year 6) was able to appreciate that forces were pushing back on her hand and also on the table below the spring.

Figure 5.25

The forces are bouncing back onto your hands from the spring when you push the spring inwards with both hands. Also, there is a force from your hands going into the spring.

The forces from your hand are pushing down onto the spring but most of the spring is pushing back. The end of the spring is pushing down onto the table.

Y6 G M

Daniel reveals an increased appreciation of the reaction forces. (See Figure 5.26). His drawing shows he can conceptualise and represent several pairs of reaction forces. His complex description of the forces together with the drawing indicates the accessibility of these ideas to some pupils at Key Stage 2.

Figure 5.26

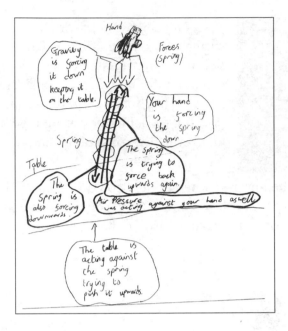

Y6 B H

A third activity was for children to sit on a chair, with and without a foam cushion. They were then asked to consider what forces were operating. Children interpreted this activity in different ways. Some described sitting on the chair in terms of rearrangement of the contents of the cushion:

When you sit on the cushion the beans go to the sides.

Others, like Sara (see Figure 5.27) identified one force, gravity, pulling down and suggested the chair kept a person up. Laura (see Figure 5.28) suggested two forces: gravity pulling down and the chair pushing up. On its own, this activity was less useful than the spring in helping children to access the idea of reaction forces. It may be that the spring provided a more direct sensation of something pushing back than did sitting on the chair. The everyday experience of the chair as providing support may have also led to a failure to appreciate the possibility of the chair pushing up.

Figure 5.27   **Figure 5.28**

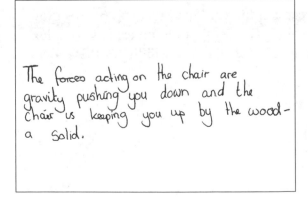

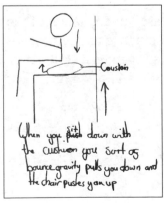

Y6 G M

Figure 5.29

The mattress spring compression activity and pushing off from a wall while seated in a wheeled box provided useful bridging analogies to understanding reaction forces. When linked to the activity of sitting on a chair, children could be helped by their teacher to make links in their understanding of the three events. An important aspect of these three practical experiences was the teacher's support in helping children to interpret the events in terms of forces. The teacher moved around the groups posing questions and challenges as new intermediate understandings were expressed. Sometimes, these questions focused on the abstract representations of the forces involved in the form of children's drawings. (See Figure 5.29). At other times, questions focused on the more tangible and intuitive aspects of the experiences, such as whether children could feel anything pushing back. Discussions tried to help children to make conceptual links between the three activities. For instance, interactions around the spring would encourage children to consider whether pressing down on the spring was similar in some way to sitting on the foam cushion. Direct links were made in these interactions between the push up of the spring and the possibility of the cushion pushing up. In each case, children were being encouraged by the teacher to make links and re-think existing representations. The extent to which children made these links was illustrated in one child's response which records the suggestion that the spring is something like the elastic band. Children's drawings of their ideas reveals insights into their developing appreciation of pairs of forces. An important aspect of promoting conceptual development was, according to the teacher, helping children to label their sensory experiences as 'reaction forces'. With some hesitation, she introduced the term to some children anticipating that they understood the concept.

> *I hope it's right I've told them it's called reaction forces. They seemed to understand the concept so I gave them the words.*

Year 6 teacher

The link between language development and conceptual development in science was explicit in the practice of some of the teachers. The comment above shows the teacher's sensitivity to the development of personal understandings. During her interactions with pupils, the teacher is searching for evidence of conceptual understandings and trying to determine the point at which it is important to label such new understandings with the correct technical words.

5.8 Helping children to develop their understanding of multiple forces.

During the elicitation, children at Key Stages 2 and 3 tended to over-estimate the number of forces they represented as acting on an object. This was the result of the tendency to nominate the *objects* which they associated with forces rather than the abstraction of the forces themselves. The effects of each force was most often considered separately with few children recognising that forces could be understood in combination. Although some children at

62

Key Stage 3 used the terms 'unbalanced' and 'balanced' forces, they had little appreciation of the implications of these terms. For example, they could not relate balanced and unbalanced forces to their differing effects on movement. Intervention aimed to help children firstly, to gain more understanding of the forces acting on objects: secondly, to help them understand that forces have a combined effect on the movement of objects. Intervention in this area was mainly approached in Key Stage 2 classrooms.

A Year 3/4 teacher used intervention to explore the effects of various forces separately. She asked children to explain how gravitational force, friction, air resistance and reaction forces work and then record which forces would act on a child on a moving bicycle. Children were encouraged to use arrows to show the direction of each force and to label each arrow. Individual drawings provided a focus for group discussions of their expressed ideas. This strategy revealed areas where children were making some progress and some points of difficulty. The work below illustrated as Figure 5.30 exemplifies an idea held by some children, that air resistance keeps things in the air. The drawing shows the deliberately constrained selection of forces nominated by the teacher for the purposes of this intervention.

Figure 5.30

<u>Multiple Forces</u>

Can you explain what these forces do?

Gravity   keeps everything on the ground.

Friction   Stops things moveing

Air resistance   it keep thing in the Air.

Reaction force   things pushing back.

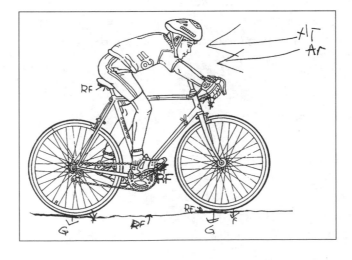

Y4 B M

In one Year 6 class, children regarded as able by the teacher addressed the effects of combined forces on movement. The children investigated the forces on a block moving at a steady speed and forces on a block moving at increasing speed. The children attached a newton-meter to the block to quantify their results. Children's work shows the interest they had in this activity. Many reported results systematically. Children varied in their ability to relate the results of their investigations to the concept of balanced and unbalanced forces. Some described their investigations but offered little in terms of interpretation.

At Key Stage 3, intervention focused on the effects of forces in combination. Children were encouraged to add and subtract forces. One teacher's journal comment suggests that children were working at the limits of their ability with only a few grasping the idea of resultant forces.

A few managed to add, subtract and obtain resultant forces.

Year 7 teacher

5.9 *Some general issues arising from the intervention experience.*

As well as the specific and precisely targeted issues which have been discussed in the earlier sections of this chapter, there are some more general points which arise from a review of intervention activities.

The whole programme of collaborative research into the learning and teaching of forces was exploratory in nature. One consequence was that all activities required careful planning, perhaps with very little precedent to build upon. Such planning for first time usage is much more time-consuming than the implementation of well-honed familiar activities where the spread of possible pupil outcomes is familiar. Teachers valued the suggestions which emerged as the result of group discussion and reflection, but only a fraction of possible activities were actually put into practice. Interventions were prioritised to address just some of the areas in which children might be helped to make progress. Consequently, the point made elsewhere is re-emphasised: the outcomes in terms of pupils' developing understanding resulting from the activities described in this chapter could not be described as optimal. Only a sub-set of ideas was addressed, with a small selection of the activities, conducted without prior experience. The next time around, teachers will be able to call on these experiences, pupils' expressed ideas will be less surprising and judgements about matching intervention activities to pupils' developmental needs will be made with greater confidence.

The idea of progression in understanding is incompatible with a dichotomous view that pupils either know, or do not know, the science. Progression in understanding, by definition, must include an acceptance of at least three points in understanding to accommodate something between complete ignorance and complete knowledge. The term we use for understandings in this middle range is 'intermediate understanding' and descriptions of progression identify possible sequences which pupils move through in learning pathways or 'trajectories'. At the project intervention meeting, teachers were concerned about their lack of overview, both of where the science was leading and of pupils' developing understanding. Such subject knowledge and knowledge of typical progression in pupils' understanding combine to inform 'pedagogical subject knowledge'. This chapter specifically, and the report generally, do not claim to present the last word on this subject. However, it is suggested that the reported activities are necessary and represent movement in the direction of a researched curriculum.

Much more could be said about the role of language in the development of pupils' conceptual understanding, but this is not the place to do so. Instances of teachers checking children's conceptual understanding before offering the correct technical label for an idea must suffice to make the point that the role of language in the socio-cultural transmission of science ideas is critical. Children's own circumlocutions and vernacular expressions are adequate until they outstrip their conceptual needs. The use of the terms 'push' and 'pull', for example, are sufficient to label a great deal of useful groundwork prior to the introduction of the abstraction of 'force' as a general label.

The rationale of validating understanding of science ideas by reference to evidence was one that seemed to gain acceptance amongst the participating teachers. Occasionally, there is a tension between this view and the pressure which insufficient time imposes on teachers. The limited time available in the intervention phases meant that there was little opportunity for lengthy interventions. There was evidence of attempts to link the evidence from several contexts and also to share a range of ideas through the medium of class discussion.

The university-based researchers were interested in the range of different representational forms which teachers would encourage children to use, since the theoretical underpinning of the model of conceptual change was one of Representational Redescription, (Karmiloff--Smith, 1992). This theoretical position suggests that pupils represent their knowledge in a variety of ways, with the central co-ordinating modality of language 'redescribing' that knowledge in a conscious, self-aware or metacognitive manner. Knowing about forces thus might entail whole-body kinaesthetic knowing, as for example, knowing intuitively about balance, or co-ordinating velocities of vehicles moving in different directions when crossing the road safely. Quantification is another mode of knowing, as is two-dimensional drawing using arrow conventions or narrative sequences which describe causal chains. The formalism of scientific investigation in which variables are defined, controlled and measured adds another dimension. Knowledge of forces might entail the representation of several such sources of internalised information which is co-ordinated into a coherent, multi-faceted account which give us that confident sense which we call 'understanding'. The important point is that this theoretical stance was entirely compatible with teachers' inclination to adopting a range of experiences using different perspectives to sharpen or bolster children's knowledge.

The next chapter documents in more quantitative form the shifts in understanding about forces which have been introduced in more descriptive terms in this chapter.

6. CHANGES IN CHILDREN'S IDEAS

This chapter presents and interprets data summarising shifts in children's ideas about forces. The data are drawn from written classroom-based assessments, most of which had been associated with physical events and phenomena managed in their classrooms by teachers during earlier phases of the project. Children recorded their ideas in booklets which contained the stimulus questions and relevant illustrations. Every child in each participating class completed the booklets. (In one or two cases, where teachers of younger children experienced time pressures, not all children completed the booklets.) The same sub-sample as was interviewed prior to the intervention activities was re-interviewed. This sub-sample comprised, as far as possible, six children from each participating class, three boys and three girls, one each of high, medium and low overall scholastic achievement according to their teacher's judgement. (This ideal distribution was adjusted and roughly balanced across the sample to accommodate the participation of single-sex schools or small schools in which the representation of pupils was particularly uneven.) The interview sample was 42 children at Key Stage 1 (5-7 years); 29 at Lower Key Stage 2 (7-9 years); 29 at Upper Key Stage 2 (10-11 years) and 18 at Key Stage 3 (11-14 years).

Each pupil was interviewed with his or her responses in the booklet being the initial stimulus for discussion. The booklet provided a focused agenda and the interview offered the opportunity to clarify the ideas already presented, as well as a chance to probe further. The minimal outcome was a clearer understanding of each child's thinking with respect to the ideas which had been targeted. In some cases, children were unable to offer any well-developed views. This was accepted in the knowledge that the requirement to make a written response was not the hurdle. In other cases, children actually changed their ideas significantly, or in emphasis, in their interview responses as compared with their written expression. When this happened, the more recent idea was recorded and coded. It was also possible for children to articulate their ideas more fully, perhaps by elaborating or providing instances or evidence. These more extensive comments provide a rich qualitative resource to illustrate and validate the more compact written responses.

All interviews were recorded in note form, using available space in the pupils' booklets. They were also audio-recorded and some were video-recorded. The latter occurred when opportunity arose, depending on the level of ambient noise and available space. All interviews were conducted during researchers' pre-arranged visits to schools, children being withdrawn singly to a quiet room or corner in the school for this purpose.

In order to make the mass of data collected comprehensible to readers, the initial discussion is presented in three sections. Any such division is to some degree arbitrary. Our decision was based on the notion of children's starting points and increasing complexity of ideas. The three sections are thus:

6.1   Some general ideas about forces;

6.2   Ideas about some specific forces;

6.3   Balanced and unbalanced forces.

There may be an understandable temptation to think of the data presented in this chapter as the outcome of a precisely targeted intervention, in the manner of an experimental treatment.

The data are emphatically not the outcome of such an experimental design. It is extremely difficult, if not impossible, for practical purposes, to match learning outcomes to the kinds of intervention to which individual children were exposed since most activities were conducted as normal classroom interactions with no outside observer or other means of recording detail. Also, from a practical perspective, we know that every teacher was not able to review children's pre-intervention responses thoroughly and match the suggested repertoire of intervention techniques to the particular ideas prevailing in their classrooms in the time available. The virtue of ecological validity – working with real teachers in their classrooms – also has its downside. The demands of the National Curriculum on teachers' time remained during this piece of research.

For the reasons suggested above, the data can be expected to be conservative in the extent to which they capture the possibilities of conceptual change. A more precise targeting of interventions to expressed ideas might be assumed to be more likely to result in optimal levels of conceptual change. What is reported here are the far more amorphous shifts within a group comprising pupils of different ages who have been subjected to a range of experiences which cannot be precisely reconstructed or reported. Nonetheless, there are shifts to be seen. Such shifts are informative, pointing to susceptibilities to development within the age group studied which might be more widely exploited.

6.1   Some general ideas about forces

The preliminary explorations with children gave indications that for many of them the scientific meaning of the word 'force' was not well understood. In consequence, questions which included the word 'force' led to responses of severely limited validity. It therefore seemed more appropriate to explore and assess children's understandings of the effects of forces through the deliberate use of the words 'push' and 'pull' within some of the intervention activities and post-intervention questioning.

6.1.1   Defining pushes and pulls

Children were asked to, 'Write how you decide whether you are pushing something or pulling it'. 'Push' is an everyday word in most children's vocabulary from an early age. The common meaning of the word is to exert a force on a body to move it away from the self (or from whatever else it is that exerts the force). At Key Stage 1 and 2, roughly twice as many children described a push as a movement forward as referred to a movement away. (See Table 6.1)

A push is when I *push a draw*

Interview response: *Pushing is when it goes forward (push a drawer).*

A pull is when I *pull my shoes on*

Interview response: *Pulling is when it goes backwards (pull my shoes on).*

Y2 B M

This pattern was reversed at Key Stage 3, the notion of movement away from the self predominating. About one fifth of Key Stage 3 subjects also appreciated that the arm could be extended to provide a push on an object *towards* the self. A small number of children (two KS1 pupils and one KS2) confused push with pull. This was not the more sophisticated realisation described by Key Stage 3 children. These younger children used a pulling movement to define a push.

Most younger children defined a pull as an action which caused a movement *backwards,* with far fewer describing it as causing a movement *towards* the self. As with responses to push, this pattern was reversed in the Key Stage 3 responses where the *towards* the self response predominated. Again, about one fifth revealed the appreciation that the arm could be extended and an object could be pulled away from the self. One Key Stage 1 child demonstrated a clear confusion of pull with push.

Table 6.1   Distinguishing between a push and a pull

	Post-Intervention			
	KS1 n=42	LKS2 n=29	UKS2 n=29	KS3 n=18
Push				
correct position of agent	-	-	-	22 (4)
forwards	43* (18)	45 (13)	59 (17)	28 (5)
away	21 (9)	31 (9)	24 (7)	44 (8)
backwards/towards	5 (2)	3 (1)	-	-
other	31 (13)	21 (6)	17 (5)	6 (1)
Pull				
correct position of agent	-	-	-	22 (4)
backwards	48 (20)	48 (14)	55 (16)	28 (5)
towards	24 (10)	31 (9)	38 (11)	44 (8)
forwards/away	2 (1)	-	-	-
other	26 (11)	21 (6)	7 (2)	6 (1)

* Pecentages, raw numbers in brackets.

68

In summary, most children understand the distinction between push and pull actions which cause movement, though a few of the younger children in the sample confused the two. Mostly, this understanding is egocentric, being defined as something done by people in relation to the position of the self. (An instance was observed amongst KS1 children of confusion that it was possible for one object – a trolley - to be moved by two people, one pushing and one pulling.) About a fifth of Key Stage 3 pupils appreciated that a push could be towards the self and that a pull could be away from the self.

6.1.2   Examples of pushes and pull

Key Stage 1 children were asked to think about what they do when they push and pull and to then write or draw four things they do which are pushes and four things they do which are pulls.

The pattern of responses was very similar for push and pull examples. About two thirds of the Key Stage 1 interview sample were able to provide four examples, as requested. Only one child was unable to give any examples. The remaining third of the sample was able to provide three, two or one examples of a push or pull.

Figure 6.1   Examples of pushes.

Y1 G H

This example drawn from the Year 1 sample is interesting because it shows a child's recognition that actions such as kicking and throwing a ball can be understood as pushes.

Figure 6.2   Examples of pulls.

my dog is pulling me across the road	*Sophie is pulling her bed sheet off*
Pulling my bike because it is not working	*I'm putting my dolly into the air and pulling it down*

RGM

Confusions between pushes and pulls were in evidence in about one fifth of the sample. This might prompt the question whether it is important for young children to be able to distinguish between pushes and pulls. In their early dealings with forces, we want children to begin to think about and understand some fundamental causal relationships. Specifically, we want them to understand that if something changes its movement, it does so *only* as the result of a force acting, very often the result of some other object moving. (Forces acting at a distance - magnetism and gravity, for example - also have to be considered at some later point.)

The verb 'to move' confuses the picture to some degree, because it is used both transitively and intransitively. We can say, 'I move the object' or, 'I move', meaning 'I move myself'. As we will see, children's ideas of movement tend to be, initially at least, preoccupied with their own and others' *intentional* movements, (using the intransitive sense of 'to move'). This has to be taken into account in the context of our efforts to encourage children to begin to think analytically about the nature of the relationship between 'bodies' in the sense in which physicists use the term. This includes distinguishing between pushes and pulls and recognising that forces occur not in isolation but between at least two objects. The issue of who or what is doing the pushing and pulling is a theme which recurs over time in the study of forces as well as in other areas of the National Curriculum. In the part of the curriculum which deals with *Materials and their Properties*, for example, the properties of materials as capable of being deformed by a push (squeezing) or a pull (stretching) are considered. Such ideas may be seen as moving the ideas of push and pull from the subjective to the objective.

6.1.3   *Examples of one push and one pull by non-human agents.*

Young children seem predisposed to be interested in active involvement, especially whole-body movement. This characteristic is often exploited by teachers. Indeed, the question requiring children to distinguish between pushing and pulling was framed in terms of children's own actions. The question thus arose as to the extent to which children were able to think of pushes and pulls as something which might happen between non-human or inanimate agents. To address this issue, children were asked directly, 'Can you think of a push that is not done by a person?', followed by a parallel question relating to pulls.

Table 6.2   **One example of a push by a non-human agent.**

	Post-Intervention			
	KS1 n=42	LKS2 n=29	UKS2 n=29	KS3 n=18
Living Things				
person	10 (4)	3 (1)	-	-
non-human animal	21 (9)	17 (5)	7 (2)	-
plant	2 (1)	-	-	-
Natural events				
wind	21 (9)	34 (10)	28 (8)	11 (2)
other natural phenomena	2 (1)	7 (2)	-	-
Human artifacts				
wheeled vehicle	14 (6)	10 (3)	17 (5)	6 (1)
mechanical	12 (5)	10 (3)	14 (4)	11 (2)
Specific named forces				
friction	-	3 (1)	3 (1)	11 (2)
upthrust	-	-	1 (3)	6 (1)
magnetic repulsion	-	-	-	17 (3)
air/resistance	-	-	3 (1)	22 (4)
gravity	-	3 (1)	3 (1)	-
Don't know,' other responses or non-instances	17 (7)	10 (3)	14 (4)	17 (3)

Table 6.2 summarises responses. It must be noted that children were required to provide only one example, so the the distribution of responses across the various response categories might be thought of as an indication of what first comes to children's minds when asked to think of a push which is not done by a person.

The proportion of children across all age groups who failed to provide a response meeting the criterion of a push by a non-human agent was higher than expected, ranging from 10 to 17 per cent. This does not include those for whom it appears to have been irresistible (five of the younger children in the sample) to include a human agent in their example, despite the clear restriction in the request. Younger children were more inclined to include examples involving animals. One example referred to a plant.

A flower pushing itself up.

Y2 G M

An animal can push a box with its head.

Y3 B H

My budgie gets its ball with its beak and pulls it up

Y1 G H

Natural events, especially the wind, were identified as capable of exerting a push by more than a quarter of children in the three younger age groups, being the predominant form of response of the Lower Key Stage 2 group. Water and geological events were mentioned by a very small minority.

When the wind blows the branches of the trees.

Y3 B H

A rock pushing another rock. It might happen if it's dead rainy.
Thunder and lightning would hit a rock and move another rock.

Y2 B L

Human artifacts were mentioned by between one fifth and one third of children in each of the three lower age groups. This was the most common category of response amongst the Upper Key Stage 2 group. Examples in this category showed roughly equal representation of wheeled vehicles and other forms of machinery or mechanical devices.

A car can force itself forwards. The engine pushes the car.

Y4 B H

We have to bear in mind that such expressions as 'the engine pushes the car' are only part of the story, perhaps sufficient to convey meaning in an everyday sense. In physics language, the story has to continue, to describe the role of the road in pushing the car, provided the wheels are turning. Such issues will be considered more fully in Section 6.3 of this chapter.

Specific named forces were the most common form of response offered by the Key Stage 3 pupils, with just a sprinkling from Key Stage 2 and no such offerings from Key Stage 1. As well as a growing technical awareness, the older pupils were no doubt more conscious of the scientific context of the probe and follow-up interviews and this possibly steered them towards a more formal exemplification. The distribution of the most frequently named forces is recorded in Table 6.2.

Air resistance

Y9 B L

In the light of these data, a tentative progression in children's thinking about pushing agents is suggested. Key Stage 1 children think of themselves and others as capable of pushing, and this is generalised to other living things. They seem to favour the intransitive use of 'to move' with its connotations of 'capable of movement', an attribute of living things. Lower Key Stage 2 children reveal more awareness of pushes happening in the natural world, especially associations with the pushing of the wind. There are strong reminders of Piaget's (1929) and Laurendau and Pinard's (1962) descriptions of children's animistic beliefs about the wind here. Upper Key Stage 2 children's responses are weighted more towards machines. We might see this as a move from subjective to objective. At Key Stage 3, the examples of pushes are abstracted and depersonalised, using technical terms for forces, in particular 'air resistance', 'magnetic repulsion', 'friction' and 'upthrust'.

The picture emerging from the parallel data concerned with identifying one pull has many similarities with the above discussion with the exception that natural events diminished dramatically. This diminution in citing of natural events is chiefly attributable to the absence of references to wind, as compared with the push examples. A small number cited wind exerting a pull, (three children at KS1 and two at KS2.) As with pushes, the youngest age group referred most frequently to pulls by living things. At Lower Key Stage 2, since natural events were scarcely mentioned, living things continued to predominate in the examples offered. The Upper Key Stage 2 children referred more to human artifacts than to any of the other response categories. (Examples of human artifacts were offered only by KS1 and KS2 pupils, with wheeled vehicles and other forms of machinery once again mentioned in equal numbers.) Key Stage 3 pupils favoured technical examples once again, though gravity (being offered by two thirds of this age group) tended to overwhelm any other possibility. It was interesting to note that gravity was also suggested by about one fifth of Upper Key Stage 2 pupils and about half as many at Lower Key Stage 2. The shift towards more formal and abstract instances of forces is in evidence at Key Stage 2, no doubt as the result of formal instruction.

In general, the age-related shift from egocentric examples through other living things, to examples in the physical world, human artifacts and specific named abstract forces is confirmed. This progression might be a useful way of structuring teaching experiences and reviewing children's learning outcomes.

Table 6.3   One example of a pull by a non-human agent.

	Post-Intervention			
	KS1 n=42	LKS2 n=29	UKS2 n=29	KS3 n=18
Living things				
person	5 (2)	3 (1)	-	-
non-human animal	31 (13)	34 (10)	17 (5)	6 (1)
Human artifacts				
wheeled vehicle	12 (5)	10 (3)	17 (5)	-
mechanical	12 (5)	14 (4)	17 (5)	-
Natural events				
wind	7 (3)	3 (1)	3 (1)	-
other natural phenomena	2 (1)	-	-	-
Specific named forces				
gravity	-	14 (4)	21 (6)	67 (12)
friction	-	-	7 (2)	-
magnetic attraction	-	3 (1)	3 (1)	6 (1)
'Don't know,' other responses or non-instances	33 (14)	17 (5)	14 (4)	22 (4)

6.1.4   'Force' as the term for pushes and pulls.

Children were asked, 'What name do we use in science for all kinds of pushes and pulls?'. The idea behind this question was to discern at what point (and how extensively) the ideas of pushes and pulls were abstracted into the more generalised understanding to which we assign the label 'force'. Table 6.4 shows that older children, in particular, responded with a range of technical words (for example, 'gravity,' 'friction,' 'power'). At Key Stage 1, only one fifth of the sample produced the word 'force'. At Key Stage 2, around half the sample was able to make the generalisation while at Key Stage 3 the figure rose to about 90 per cent. The following responses illustrate the elaborations offered by two children who used the word 'force'.

Int   *What does a force do?*

Ch   *It makes things move.*

Int   *Does it do anything else?*

Ch   *Makes it move more. It can make it stop.*

Ch   *[Forces] change the nature of things around them. Gravity keeps us
            down. Friction slows us down. Magnetic forces attract some metal
            objects.*

Table 6.4   'Force' as the term for pushes and pulls

| | Post-Intervention | | | |
	KS1 n=42	LKS2 n=29	UKS2 n=29	KS3 n=18
force	21 (9)	59 (17)	48 (14)	89 (16)
other technical names	2 (1)	14 (4)	45 (13)	11 (2)
word related to force	2 (1)	3 (1)	-	-
description of action	-	3 (1)	-	-
Don't know,' and other responses	74 (31)	21 (6)	7 (2)	-

This finding underlines the age-appropriateness of dealing with specific events to be referred to as 'pushes' and 'pulls' at Key Stage 1. It became apparent as the result of talking to young children that the term 'force' actually constitutes a difficult abstraction, one that is not easily accessible to children at Key Stage 1.

6.1.5   *Pushes and pulls in order of magnitude.*

Quantification is central to scientific enquiry. In the course of the research, consideration was given to the manner in which children might be encouraged to think about forces as phenomena having magnitude. The logico-mathematical operations explored by Piaget (1929) offered a useful precedent, especially when coupled with the practices, commonly adopted in Key Stage 1 and Key Stage 2 classrooms, used to introduce other concepts of measurement. Putting examples in order of magnitude was one of the starting points decided upon. Children were asked to think of a very small push, then a very big push and finally a medium-sized push, writing or drawing their choices for each in the boxes provided. Results are summarised in Table 6.5. A parallel set of questions was presented with respect to pulls and the results of that aspect of the enquiry are presented in Table 6.6.

Table 6.5   Examples of pushes in order of magnitude.

| | Post Intervention | | | |
	KS1 n=42	LKS2 n=29	UKS2 n=29	KS3 n=18
3 pushes, ordinal	45 (19)	55 (16)	34 (10)	28 (5)
3 pushes, not ordinal	36 (15)	31 (9)	38 (11)	22 (4)
pushes and pulls	5 (2)	-	-	11 (2)
ambiguous responses	2 (1)	3 (1)	3 (1)	22 (4)
2 pulls, ordinal	7 (3)	10 (3)	10 (3)	6 (1)
other responses	5 (2)	-	14 (4)	11 (2)

Table 6.6   Examples of pulls in order of magnitude

| | Post-Intervention | | | |
	KS1 n=42	LKS2 n=29	UKS2 n=29	KS3 n=18
3 pulls, ordinal	38 (16)	59 (17)	31 (9)	33 (6)
3 pulls, not ordinal	33 (14)	21 (6)	21 (6)	11 (2)
pulls and pushes	2 (1)	7 (2)	7 (2)	-
ambiguous responses	5 (2)	7 (2)	17 (5)	39 (7)
2 pulls, ordinal	10 (4)	3 (1)	10 (3)	11 (2)
other responses	12 (5)	3 (1)	14 (4)	6 (1)

76

Results show some surprising features which might lead one to question the reliability of the data were it not for the similarities between the two sets of results. There is not a trend of increasing success with age. In both instances, the two younger groups perform better than the two older ones, with the peak of performance at Lower Key Stage 2. (See Figure 6.3).

Figure 6.3   Pulls and pushes in order of magnitude

	very big	**medium sized**	**very small**
pull	*pulling a wagon across the road*	*pulling a toy*	*pulling a a leaf off a tree* 
push	*pushing a lampost over* 	*pushing my dad out of bed*	*pushing a teddy over*

Y2 G H

There is a very high incidence in all groups of the three examples given being valid cases of pushes or pulls but not ordinal in magnitude. This tendency was slightly less amongst the Key Stage 3 respondents who were more likely to offer ambiguous responses. That is, the order of magnitude of the pushes or pulls which they suggested could not be judged by an independent assessor. This was due to a tendency to include named forces of unspecified magnitude. (See Figure 6.4).

Figure 6.4   Pulls and pushes in order of magnitude

	very big	**medium sized**	**very small**
pull	gravity	person	apple
push	rocket	person	piston

Y7 G H

A significant proportion of pupils offered only two instances of pushes and two instances of pulls: eight per cent of the total sample in each case, while a similar proportion offered other responses not meeting the questions' criteria.

How might these patterns of success (and lack of success) be explained? There is no doubting that the older pupils knew more about forces in a formal sense, for this is confirmed by their responses to many of the other questions posed. On this occasion, it seems likely that the Lower Key Stage 2 children responded in an intuitive manner and succeeded in making unambiguous qualitative distinctions. It is possible that knowing more about forces interfered with the Key Stage 3 pupils' willingness to respond in a similarly direct qualitative fashion. Responding by reference to named forces required that differences in magnitude needed to be equally explicit, in formal terms. These tended not to be explicit, so many of the Key Stage 3 responses could not be said to describe an unambiguous ordinal relationship in the values suggested.

6.1.6   *Identification of a force-meter and understanding its uses*

A consideration of forces in the curriculum would be expected, at some point, to deal with the issue of how force is quantified and measured. In the course of the research work with teachers, the team came to realise the part that children's own non-standard measuring devices might play. The construction and use of such devices was more developed as an idea rather than an actuality in the research schedule, but it was of interest to ascertain the extent of the sample's familiarity with standard devices for measuring force, namely force-meters (or newton-meters). The manner in which familiarity with such devices was tapped was through the use of a photograph accompanied by the question, 'What is this measurer called?' Responses to this question before and after intervention are summarised in Table 6.7. (Since use of force-meters at KS1 was not expected, the question was not presented to the youngest age group.)

Figure 6.5

Post-intervention, about half of the Lower Key Stage 2 sample named the pictured measuring device as a newton-meter or force-meter.

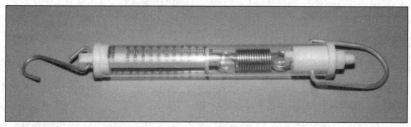

This represented an increase in comparison with the same sample's responses pre-intervention. Of course, a number of children referred to the device as a 'spring balance', which it undeniably is, often being graduated in grams as well as in newtons.

Table 6.7   Identification of a newton-meter

	Pre-Intervention				Post-Intervention			
	KS1 n.a.	LKS2 n=29	UKS2 n=29	KS3 n=18	KS1 n.a.	LKS2 n=29	UKS2 n=29	KS3 n=18
newton-meter		24 (7)	37 (11)	56 (10)		45 (13)	34 (10)	61 (11)
force-meter		-	-	-		7 (2)	-	17 (3)
pull-meter		-	-	-		3 (1)	-	-
spring balance		14 (4)	-	22 (4)		10 (3)	-	17 (1)
weigher		14 (4)	-	6 (1)		-	-	-
other named instrument		10 (3)	7 (2)	-		3 (1)	7 (2)	-
'Don't know,' and other responses		37 (11)	55 (16)	17 (3)		31 (9)	59 (17)	6 (1)

The question following the identification of the force-meter asked children, 'What does it measure?'. The most generalised and accurate response might be expected to take the form, 'It measures force', or 'It measures force in newtons'. Since newton-meters often double as devices for weighing things, responses referring to 'weight' or units of mass were also expected. Responses are summarised in Table 6.8.

Pre-intervention, younger pupils' responses referred predominantly to 'weight' or 'grams' with none mentioning 'force' and only a few mentioning 'newtons'. From the Key Stage 3 pupils (who would be expected to have had some experience of measuring forces) about one quarter of responses referred to 'force', one quarter to 'newtons' and one quarter to 'weight'.

The post-intervention data strongly suggest that some modification of ideas took place. 'Force' or 'newtons' accounted for about one quarter of the Lower Key Stage 2 responses, about one third of the Upper Key Stage 2 responses and three quarters of those at Key Stage 3. Responses referring to the measurement of 'force' arguably suggest awareness of a continuous physical quantity whereas the specification of units might connote a more limited understanding of discrete categories.

Table 6.8   **Property measured by newton-meter**

	Pre-Intervention				Post-Intervention			
	KS1 n.a	LKS2 n=29	UKS2 n=29	KS3 n=18	KS1 n.a.	LKS2 n=29	UKS2 n=29	KS3 n=18
force	-	-	28 (5)		10 (3)	14 (4)	50 (9)	
pulls	3 (1)	-	6 (1)		10 (3)	3 (1)	11 (2)	
weight	48 (14)	24 (7)	28 (5)		28 (8)	21 (6)	6 (1)	
other named force	-	-	6 (1)		3 (1)	-	-	
newtons	3 (1)	21 (6)	28 (5)		14 (4)	21 (6)	28 (5)	
grams	10 (3)	17 (5)	-		7 (2)	3 (1)	-	
mass	-	-	-		-	-	6 (1)	
named objects	21 (6)	14 (4)	-		14 (4)	10 (3)	-	
other physical quantities	7 (2)	3 (1)	-		7 (2)	10 (3)	-	
Don't know' and other responses	7 (2)	21 (6)	6 (1)		7 (2)	17 (5)	-	

Pupils were further asked to, 'Give two different ways that it can be used to measure'. The idea behind this request was to point up the distinction between horizontal and vertical uses of the instrument. Those pupils who thought in terms of measuring 'weight' would be unlikely to do this with the meter in the horizontal position. The question thus offered a strong invitation to express ideas about measuring forces more generally, since this would be the most likely reason to use the meter in the horizontal orientation. Table 6.9 summarises the results of this part of the enquiry.

Pre-intervention, more than half of the students at both Key Stage 2 and Key Stage 3 suggested a *vertical* use of *hanging things* on the meter or *pulling down* on it in the vertical position. Many referred to the possibility of using the meter to weigh objects. Post-intervention, references to vertical use of the meter actually diminished.

You can put it on something and lift it up and see what it weighs. You could try and lift
        sugar.
If it goes down to 10 you've got loads in it.

                                                                            Y3 B M

You can put something on the hook and it will pull down and it will measure.

Y4 G L

You put weights on the hook and hold it and read the newtons.

Y9 B H

You pull it from the end. You can measure how big the pulling force is.

Y9 B H

Table 6.9   Using newton-meter vertically and horizontally

	Pre-Intervention				Post-Intervention			
	KS1 n.a.	LKS2 n=29	UKS2 n=29	KS3 n=18	KS1 n.a.	LKS2 n=29	UKS2 n=29	KS3 n=18
Indication of vertical use								
hang/pull down		55 (16)	55 (16)	61 (11)		34 (10)	41 (12)	28 (5)
weight		28 (8)	10 (3)	33 (6)		17 (5)	14 (4)	33 (6)
Indication of horizontal use								
pull along		3 (1)	3 (1)	28 (5)		10 (3)	34 (10)	17 (3)
Summary of responses								
vertical and horizontal uses suggested		3 (1)	3 (1)	28 (5)		7 (2)	28 (8)	17 (3)
vertical use only		79 (23)	66 (19)	67 (12)		45 (13)	32 (9)	44 (8)
horizontal use only		-	-	-		3 (1)	3 (1)	-
other responses		17 (5)	31 (9)	6 (1)		45 (13)	38 (11)	39 (7)

A suggestion for the horizontal use of the force-meter was comparatively rare amongst Key Stage 2 pupils pre-intervention, while a quarter of Key Stage 3 subjects indicated that it could be used in the horizontal position to *pull* things. This awareness increased in the Key Stage 2 sample following intervention activities, but diminished amongst those at Key Stage 3.

It seems that the use of the force-meter for measuring forces was not at all well established or understood in any of the age groups participating in the study. This may be attributable to a lack of experience. This situation is not helped by the fact that there are few commercially available force-meters which have been designed to measure pushes. It is tempting to suggest that Key Stage 3 pupils' uncertainties were actually increased as the result of further activities, perhaps their confidence about its use as an instrument for weighing things having been put on hold.

Table 6.10 indicates that, in contrast to their lack of knowledge about how to use a force-meter and for what purpose, many pupils could name the unit in which the force-meter measures. About one fifth of Key Stage 2 pupils hazarded the guess that the meter measured in units of length, possibly cued by the linear scale (visible in the photograph of the meter).

Table 6.10   Unit in which newton-meter measures

	Pre-Intervention				Post-Intervention			
	KS1 n.a.	LKS2 n=29	UKS2 n=29	KS3 n=18	KS1 n.a.	LKS2 n=29	UKS2 n=29	KS3 n=18
newtons		14 (4)	34 (10)	94 (17)		48 (14)	31 (9)	100 (18)
gram/kg/mass unit		31 (9)	34 (10)	-		-	14 (4)	-
newton and gram		-	-	-		3 (1)	3 (1)	-
other units		21 (6)	3 (1)	-		21 (6)	28 (8)	-
Don't know' and other responses		34 (10)	28 (8)	6 (1)		28 (8)	24 (7)	-

6.1.7   *Representing forces using the arrow convention*

Several questions invited the use of arrows to identify and label forces acting in various situations. Responses reveal, to some extent, pupils' understanding of the use of arrow conventions to describe the direction and magnitude of forces though not all cues to use arrows made an explicit request for a formal representation. More general issues associated with the use and interpretation of arrows will also be discussed in later sections. (See for example, Section 6.3.2.1). At this point, a specific and focused requirement to *interpret* the use of given arrows is discussed. The reference is to a question showing a milk bottle held by a hand. A long arrow points vertically upwards while a short arrow points vertically downwards. Children are asked, 'What exactly do the arrows on the drawing tell you about the forces acting?'. The responses were analysed for references to *direction* and *magnitude* (*Movement* represented by the arrow convention is discussed in later sections.) While it was possible to interpret the drawing as representing a bottle decelerating towards the ground, this was not an expected response in the age group under consideration. (Another explanation which is consistent with the drawing is that the bottle is accelerating away from the ground, or more simply, moving upwards. It was the interpretation of the magnitude and direction of the arrows which was drawn from responses for the purpose of this discussion even though pupils might have volunteered information about the resultant movement).

This question was presented to the Key Stage 2 and Key Stage 3 pupils as a pre-intervention activity. The results were that only about one third of each age group referred to the direction of the force while even fewer (one pupil each at Upper KS2 and KS3) referred to magnitude. Data are summarised in Table 6.11. The inference was drawn that Key Stage 2 and Key Stage 3 children in the research sample, for whatever reason, had not been exposed to teaching of this particular convention, prior to intervention. The view was that these conventions were certainly appropriate to the Key Stage 3 sample pupils, so the question was posed a second time to this group, post-intervention, but not to the Key Stage 2 group. Judgement as to the appropriateness for Key Stage 2 was suspended.

Table 6.11   Arrow representation of forces - direction and magnitude

	Pre-Intervention				Post-Intervention			
	KS1 n.a.	LKS2 n=29	UKS2 n=29	KS3 n=18	KS1 n=n.a.	LKS2 n.a.	UKS2 n.a.	KS3 n=18
Direction								
direction, general		3 (1)	10 (3)	28 (5)				-
upward and downward force		14 (4)	7 (2)	6 (1)				83 (15)
upward force only		10 (3)	3 (1)	-				-
downward force only		3 (1)	3 (1)	-				6 (1)
direction not mentioned		69 (20)	76 (22)	67 (12)				11 (2)
Magnitude								
upward force greater		-	-	6 (1)				50 (9)
upward force same size		-	3 (1)	-				6 (1)
size not mentioned		100 (29)	97 (28)	94 (17)				44 (8)

The post-intervention responses of the Key Stage 3 group showed a dramatic positive shift in the appreciation of the directional information about forces conveyed by the arrows on the milk bottle. There was also a very significant positive shift in their appreciation of magnitude, but not quite to the same degree as the shift in understanding with respect to direction.

The down arrow is the force of gravity and the up arrow is the force of the person's hand on the bottle. The upward arrow is longer so the upward force is greater than the downward force so the bottle is being picked up or lifted.

Y9 B M

The boy is picking up the milk bottle. You can tell because the force pulling up is more than the force of gravity.

Y7 B H

Forty-four per cent of the Key Stage 3 sample did not link the size of the arrows to the magnitude of the forces on the bottle.

There is one arrow which means gravity and one arrow which means weight.

Y7 G L

The dramatic shifts apparent in the data in Table 6.11 suggest that, when directly addressed, the arrow conventions of direction and magnitude of forces seems to be fairly readily accessible to Key Stage 3 pupils. This fact, together with some Key Stage 2 teachers' success in helping children to represent forces using arrows, suggests that the convention is worth exploring with Key Stage 2 pupils also.

6.2   Ideas about some specific forces

6.2.1   Ideas about the gravitational force of the Earth

The Key Stage 2 Programmes of Study in the National Curriculum refer to the knowledge *'that objects have weight because of the gravitational attraction between them and the Earth'*. This carefully worded expression acknowledges the necessity of conceptualising gravitational attraction as something that happens *between* objects. In time, children will be expected to move to an appreciation of the role of mass and distance in such relationships, but our focus was on direction only. Asking whether the Earth's gravitational attraction is a push or a pull presupposes that it is understood as a force between two objects. This fact justifies presenting such a question but does not imply a belief on the part of the researchers that children universally shared such a view. Indeed, an important developmental issue seems to be a movement away from the personalising of forces towards an objective view. An *intermediate understanding* might be that forces are seen as the properties of objects.

While a view that gravity is a property of the Earth to attract objects to its surface would be less than correct in the physicist's understanding, in an educational and developmental context, we have to make decisions as to whether such stated beliefs constitute positive progress in the direction of a more complete conventional scientific understanding. If we believe such ideas constitute conceptual progress, we must decide how to treat them as useful while stopping short of validating them, for all time, as correct beliefs.

Children's confusion as to whether the effect of the Earth's gravitational force on objects is a push or a pull is established in the literature concerned with ideas about forces. The finding that some children might describe gravity as *both* a push and a pull adds another dimension. The push/pull confusion with respect to gravity became apparent in the research reported here during initial explorations of children's thinking. Once recognised, the issue was specifically addressed. Children were asked, 'Is the effect of gravity on objects a push, a pull or both?'. They were then asked to tick one of the three boxes matching these options, followed by the request to, 'Explain how gravity works'.

Table 6.12 summarises children's choices of the direction in which the force of gravity acts on objects. Pre-intervention data reveal about the same proportion of Lower Key Stage 2 children describing gravity as a push as described it as a pull. Both of these reduced in frequency post-intervention, with choices shifting towards 'push *and* pull'. This group may be influenced by context-specific ideas about gravitational forces which are assumed to work in different directions in different situations.

Table 6.12   Effect of gravity on objects in terms of push and pull

	Pre-Intervention				Post-Intervention			
	KS1 n.a.	LKS2 n=29	UKS2 n=29	KS3 n=18	KS1 n.a.	LKS2 n=29	UKS2 n=29	KS3 n=18
gravity is a pull		41 (12)	41 (12)	72 (13)		31 (9)	55 (16)	83 (15)
gravity is a push		45 (13)	55 (16)	22 (4)		21 (6)	24 (7)	6 (1)
gravity is a push and a pull		7 (2)	3 (1)	6 (1)		45 (13)	17 (5)	11 (2)
no response		7 (2)	-	-		3 (1)	3 (1)	-

The Upper Key Stage 2 post-intervention data show an additional 14 per cent defining the force of gravity as a pull with the 'push' responses declining. As with the younger group, the tendency to define gravity as both a push and a pull increased.

At Key Stage 3, the already high proportion of pupils defining gravity as a pull increased from 72 per cent to 83 per cent.

The post-intervention summary is that the Earth's gravitational force on objects was correctly defined as a pull by about one third at Lower Key Stage 2, about half at Upper Key Stage 2 and about four fifths at Key Stage 3. If these cross-sectional data are taken as offering a clue to progression in understanding, the development is relatively steady across the seven years under consideration. It is nonetheless the case that a significant number of children of all ages viewed the Earth's gravitational force as a push, the figure remaining at six per cent even at Key Stage 3.

The follow-up question, 'Explain how gravity works.', was designed to encourage children to elaborate their basic statement about direction towards description or explanation of the *mechanism* of the Earth's gravitational force. Results are summarised in Table 6.13.

Table 6.13   How gravity works

	Pre-Intervention				Post-Intervention			
	KS1 n.a.	LKS2 n=29	UKS2 n=29	KS3 n=18	KS1 n.a.	LKS2 n=29	UKS2 n=29	KS3 n=18
gravity is:								
attraction between masses	-	-	-		-	3 (1)	6 (1)	
force between Earth and masses	-	-	-		3 (1)	-	11 (2)	
pull of Earth	-	-	39 (7)		3 (1)	17 (5)	44 (8)	
due to spin of Earth	3 (1)	-	-		3 (1)	3 (1)	6 (1)	
keeps things down	14 (4)	10 (3)	11 (2)		21 (6)	38 (11)	28 (5)	
stops floating away	3 (1)	17 (5)	22 (4)		3 (1)	7 (2)	-	
pulls things down	-	7 (2)	6 (1)		24 (7)	17 (5)	-	
pushes things down	3 (1)	-	6 (1)		10 (3)	3 (1)	-	
caused by push of air	7 (2)	3 (1)	-		3 (1)	3 (1)	-	
because Earth like a magnet	-	-	-		3 (1)	-	6 (1)	
Don't know,' and other responses	69 (20)	62 (18)	17 (3)		24 (7)	7 (2)	-	

Before intervention activities, no children expressed an understanding of gravitational force in the most generalised manner, as an attraction *between masses*, or more specifically, as an attractive force *between* the Earth and other masses. About one third of Key Stage 3 pupils described gravity as a property of the Earth.

Gravity pulls things to the centre of the Earth

Y9 B L

Two thirds of Key Stage 2 children could not begin to explain how gravity works. Those who were able to formulate a response tended to describe gravity in terms of its effects - something which keeps things down or stops things floating away. About one third of Key Stage 3 pupils used similar descriptions to explain how gravity works.

Because gravity pulls you down so you don't float

Gravity stops things floating and keeps them on the ground.

Though we cannot specify precisely what form they took for individual pupils, it seems that intervention activities had a significant impact on children's thinking. A very small number of children reformulated gravitational force as a force *between* bodies (or more specifically, between the Earth and other bodies).

Gravity is the force which acts upon masses to pull them down to the ground. It is the the force of attraction between two masses.

The Earth attracts different masses to it. The attraction is called gravity. Heavier objects attract objects with greater force. Jupiter has more gravity than Earth.

Gravity comes from the core of the space bodies and pulls thing to it. The Moon has gravity but it is weaker.

The more limited explanation of gravity being 'the pull of the Earth' increased in popularity as did the description of it as something that 'keeps things down'.

Ch   *Gravity pushes and pulls and makes you stay on the ground.*
        Gravity is in space. No, not in space. Space has no gravity.

Int   *Where would you find gravity if you were to look for it?*
Ch   *On the Earth there is gravity that keeps us down. You find it on the floor because gravity keeps us down. No, gravity is up in the air because you stay down it's pushing you down to stay on the floor. If you bounce a ball on the floor gravity pushes it back up.*

This example and the one that follows are examples of children describing gravity as a push and a pull within the same explanation.

Ch   *Gravity is a pull. It goes in a straight line. On light people the gravitational pull is not big. On fat people the gravitational pull is big.*

Int   *How does this happen like that?*

Ch   *They have a bigger mass. More weight than a thinner person or a fat person. So there would be more weight adding, pushing down when you stand up. All your weight is down at the bottom. If you have something heavy in your pockets your pants will fall down.*

Y6 B M

gravity eminates from the centre of the Earth. Round the Earth there is a gravitational field and anything under that is pulled towards the Earth.

Y9 B M

There is evidence that children were testing some causal ideas, which is pleasing, even though these were erroneous. For example, some referred to the spin of the Earth

Ch   *The Earth spins at a very fast speed and the spinning pulls objects down. Astronauts would go to different planets. Earth has a very strong pull and the Moon has a light pull. The Earth's gravity holds the Moon where it is and it (Moon) can only move slowly because the Earth's gravity stops it moving.*

Int   *Why does the Moon have less gravity?*

Ch   *The Earth's gravity holds the moon in orbit and stops the Moon spinning fast and that's why it has less spin and less gravity.*

Y3 B H

Others talked of the push of air.

Ch   *Gravity pushes you down to be able to stay on the Earth. You can force things up but gravity pushes them back down.*

Int   *Where would you find gravity on the Earth?*
Ch   *All round us. In the air. Inside things - can. You throw the can up it will come back down because you've got gravity in the air pushing down on top of the can.*

Y6 B L

Ch   *You find gravity down here. Walking on paths, floor, ground. If it wasn't there you would just float up.*

Int   *How does it keep you on the ground?*

Ch   *Down here there is air. Smoke rises like air because there is no air up there. There is air on the floor. Gravity is just like air. There is no air in space. Air keeps you down.*

Y4 G M

The simile of the gravitational force of the Earth being like magnetic attraction was also in evidence.

Ch   *The centre of the Earth is magnetic and it pulls everything towards it. It brings anything with mass towards it. Everything that has mass has gravity on. The bigger it is, the bigger the force of gravity.*

Int   *How does gravity work then?*

Ch   *It is like a magnet. A magnet attracts other metal. The nearer the metal is to the magnet the bigger the attraction. The nearer the two things are the bigger the pull. If you get too far away they won't attract each other.*

Y7 B H

One might sympathise with children's struggles to make sense of this pervasive force which is so much part of daily life that we cease to notice it. Beyond such everyday experience we have very little direct evidence of gravitational attraction between masses. So we tend to use the word 'gravity' as shorthand to name the gravitational attraction between objects and the Earth. Taking a wider perspective we may point to the effect of the Moon's gravitational attraction on the Earth's oceans and where these forces are even greater, between planets. We have to accept that this is simply the way the universe appears to work. Those pupils who have stated that gravity is an attraction between masses (or between the Earth and other objects) have taken their thinking as far as could be hoped, and we should perhaps be delighted that seven per cent of the sample achieved this level of understanding. The fact that these five pupils included one each at Lower and Upper Key Stage 2 is reason for optimism that this important idea might be made more widely accessible.

Children's explanations as to how gravity works were also analysed for directionality, for this aspect was deemed to be an important indicator of the degree of generality (or 'depth') of their understanding. The following levels, indicating movement in understanding towards increasing generality, can be defined:

1.   If we lack a sense of living on the surface of a planet, we can still operate a definition in which *gravity causes objects to fall to the ground*. Perhaps even less abstracted, more local still, is the notion that *gravity causes objects to fall downwards*.

2.   Still fairly parochial, gravitational force can be thought of as a force acting *between the Earth and other bodies near the Earth*. An elaboration of this idea is an appreciation of the importance of the centre of the Earth as the notional point towards which objects are attracted.

3.   At its most general, gravity is a force which might be imagined to operate between two masses anywhere in the universe.

All of these ideas in their various degrees of completeness incorporate some understanding of the force of gravity. Pre-intervention, the most common outcome was no clear indication of downward attraction (see Table 6.14). Of the responses which indicated direction, the most common was the most local, least abstracted idea of 'downwards'. Next most frequent, though seen only at Key Stage 3 and even there, in small numbers, was the idea of gravity as a force of attraction towards the centre of the Earth. Least common was the most abstract level of definition of attraction between the centres of masses. This same pattern recurred post-intervention, though with far more pupils indicating the direction in which they believed the force of gravity to be acting, and large increases in the direction of scientifically more accurate (and more abstracted) responses.

Table 6.14   Direction of force due to gravity

	Pre-Intervention				Post-Intervention			
	KS1 n.a.	LKS2 n=29	UKS2 n=29	KS3 n=18	KS1 n.a.	LKS2 n=29	UKS2 n=29	KS3 n=18
1. towards centres of masses		3 (1)	-	-		-	3 (1)	6 (1)
2. towards centre of Earth		-	-	17 (3)		7 (2)	7 (2)	33 (6)
3. downwards, towards ground		21 (6)	21 (6)	56 (10)		59 (17)	72 (21)	33 (6)
No clear indication of downward attraction		76 (22)	79 (23)	28 (5)		35 (10)	17 (5)	28 (5)

Another question which probed ideas about gravity used the activity of a ball thrown up in the air. Children were asked whether gravity is acting on the ball, (i) when it is moving upwards and (ii) just when it reaches its highest point. In this instance, children have to be clear about the interaction between the two forces, that of the hand making the throw and that of gravity. They may also be aware of other forces acting, such as air resistance. The interacting forces in this context cause us to become aware of some wider aspects of children's understanding of how gravity works. For example, it becomes apparent that the Earth's gravitational force is not always treated by them as a constant force on the ball, but as a variable interaction between the force upwards exerted by the hand and the force downwards of gravity. While the ball was moving upwards, half the Lower Key Stage 2, about a third of the Upper Key Stage 2 and about a fifth of Key Stage 3 pupils thought that gravity was *not* acting. The movement upwards seems to have precluded the possibility, in their minds, that a force downwards could be acting. A small minority suggested that gravity would be acting but to a reduced extent during the ball's upward movement. In contrast, at the apex of the trajectory, all the Key Stage 3 children believed that gravity was acting, together with the majority of their Key Stage 2 counterparts.

There's another force which is stronger while it is going upwards. It's up in the air. It's always there if you push the ball upwards then the force in the air is helping the ball upwards. It's stronger than gravity. When it's at its highest point gravity and the other force are equal. Gravity will eventually pull it back down. When it went up it used its strength against gravity. When it was at its highest point they were equal. When it was coming down it has lost its strength.

Y6 G M

Ch *Because you give it a push force which makes it go up gravity is still acting when it goes up. When it gets to its highest point it just drops. The force of your hand throwing it really hard makes it go up and it just drops when the gravity gets too strong.*

Int *Why does gravity get too strong? Why is it stronger when the ball is at its highest point?*

Ch *The weight of the ball, it gets too heavy to go high any more. The speed of the ball pushes air out of the way so the air can make way for the ball. On different planets, the stronger the gravity, the ball would go to a different height. Where there is less gravity the ball would go quite high, about 25ft. Where there is no spin at all the ball would go high and may not come down. Gravity is always the same on Earth, it is always spinning. When the ball stops, gravity has a chance to pull it down.*

Y3 B H

Ch *Nothing is being pulled down because gravity is not acting. It won't go down. When it gets to its height it can't stay there because gravity pulls it down.*

Int *How does it happen to be at that height?*

Ch *Gravity is there, no matter how high, it is still there. It's not very high but as high, as you can throw something.*

Y4 B L

In another question, a car with driver and passenger were illustrated. The passenger was described as 'just letting go of the can' with her arm extended through the car's window. The further information was offered that, 'The car is being driven forwards quite fast.' Children were asked to mark the drawing to show the point where they thought the can would first hit the road. They were then asked to explain why they thought the can hit the road at the point which they had marked. Responses to this question are discussed more fully in Section 6.3. For the moment, attention will focus on the incidence with which pupils chose to make references to gravity as the force which caused the can to hit the road. Such references to gravity were, in the context of the can dropped from a moving car, rare.

Seven pupils in the two oldest age groups mentioned gravity pre-intervention, increasing to eleven pupils across the entire age span post-intervention. This total included only about one fifth of the Key Stage 3 pupils. A dropped object from a moving vehicle appears not to suggest the force of gravity. As with the vertically thrown ball discussed above, it seems that other forces which are acting on an object moving through the air readily dominate children's thinking and may be deemed to negate the effects of gravity.

A brief video clip (not at all clear in quality) of Neil Armstrong's hammer and feather experiment on the Moon was shown to all children. Aspects of children's responses are reported here since many of them invoked the concept of gravity to explain what they saw on the video-recording. Armstrong is shown holding each object at arm's length before releasing them. It was confirmed that, 'They both hit the surface of the Moon at the same time.' and children were asked to explain, 'Why does this happen on the Moon but not on Earth?'. The Key Stage 1 sample was initially included but was omitted post-intervention since they seemed to make so little sense of what was shown in the film. (The visual quality of the evidence presented in the film made acceptance of the objects and the event an act of faith.)

About one third of the entire Key Stage 2 and 3 sample, both before and after intervention, suggested that the result occurred as it did because there is no gravity on the Moon, (see Table 6.15).

Ch   *On the Earth there is gravity to pull things down like the hammer.*

Int   *What happens on the moon?*

Ch   *Moon? No gravity. On Earth heavier things move down first, but on the moon they are both the same weight. Gravity pulls things down on Earth. No gravity makes the hammer lighter.*

Y4 B L

It is difficult to understand how children are thinking to arrive at this sense which they made of the Armstrong experiment. Pre-intervention, about one third of the Key Stage 3 sample suggested that the outcome was because of *less* gravity on the Moon; this proportion almost doubled in the post-intervention interview responses. The example above shows one example of the kind of complete chain of reasoning needed to arrive at the 'no gravity' reasoning.

Table 6.15   Hammer and feather falling on Moon - role of gravity.

	Pre-Intervention				Post-Intervention			
	KS1 n=42	LKS2 n=29	UKS2 n=29	KS3 n=18	KS1 n.a.	LKS2 n=29	UKS2 n=29	KS3 n=18
no gravity on Moon	8 (3)	28 (8)	52 (15)	28 (5)		48 (14)	41 (12)	22 (4)
less gravity on Moon	-	14 (4)	17 (5)	33 (6)		17 (5)	17 (5)	61 (11)
more gravity on Moon	-	-	-	6 (1)		10 (3)	7 (2)	-
other responses	-	24 (7)	21 (6)	11 (2)		10 (3)	3 (1)	-
gravity not mentioned	92 (39)	34 (10)	10 (3)	22 (4)		14 (4)	31 (9)	17 (3)

One Lower Key Stage 2 child explained a supposed absence of gravity on the Moon as attributable to the lack of air.

The hammer is heavier than the feather but in space there is no gravity so everything would fall at the same rate. Down here the hammer would fall first because there is air so there is gravity. Gravity and air are just like brothers and sisters.

Y4 G M

About one fifth of the Key Stage 2 and Key Stage 3 samples combined explained the simultaneous impact of the hammer and feather on the Moon correctly, in terms of an absence of air resistance. There was no strong age association with this response and the frequency rose only very slightly in the post-intervention responses.

Ch   *On Earth the hammer would go first because the feather would float down.*

Int   *Why does it float like that?*

Ch   *It's light so there's an upwards force from the air.*

Y7 B M

Another question in the context of astronauts on the Moon has some relevance to this discussion of gravity. The question was posed, (cued by video footage of Moon-walking in which astronauts are seen moving gracefully if ponderously, wearing bulky suits, helmets and boots), 'Why do astronauts wear big boots when they walk around on the Moon?'. Only the Key Stage 3 sample was presented with this question before and after intervention and none answered in terms of the large boots offering grip or protection. Fifty-six per cent suggested that the boots were to give added *weight,* though without any reference to the gravitational force of the Moon.

They wear big boots to help them to stay down better so they don't float away into space.

Y9 B H

Twenty-eight per cent suggested that the boots added weight in the context of there being *less* gravity on the Moon (as compared with the Earth).

Because there is less gravity on the moon and the boots keep them down. Because it is higher up than Earth so there is less gravity.

Y7 G L

Seventeen per cent of this same Key Stage 3 group suggested that the boots added weight because of an *absence* of gravity on the Moon.

Because the is no gravity on the moon the weight of the boots pull them downwards.

Y7 G H

The line of argument seems to be that in the absence of gravity, extremely heavy boots are needed. It appears that these respondents have completely separated weight from the force of gravitational attraction and advocate 'heaviness' as a substitute for the absent gravity.

Ch   *On Earth you have weight but on the Moon you don't have weight, you only have mass. So where there is gravity, there is weight.*

Y7 B H

Perhaps insufficient attention had been paid to weight or heaviness as the result of gravitational attraction in the course of the intervention activities to which these pupils were exposed.

The distinction between measuring mass and force in appropriate units was directly addressed with Key Stage 3 pupils only. The problem was set by describing someone having a 500 gram pack of butter which, when hung on a newton-meter, gave a reading of five. The question was posed, 'Why does it read 5 and not 500?'. Half of the responses suggested that the same property was being measured, but using different units. That is, no distinction was made between mass and weight.

Because it's measured in Newtons and 1N =100g.

Y9 G M

1N is the equivalent of 100g. They are measuring something in different units.

Y7 G M

About ten per cent gave responses which included 'weight' or 'force' as the property being measured in newtons. We cannot be certain that such responses incorporate a clear distinction between weight and mass, a recurrently problematic distinction for pupils.

The force is five and it is measured in newtons. It is not measured in g, it is measuring in newtons. It is measuring the force. If it is in g it is measuring weight.

Y7 B H

Another question inviting references to gravity was that which asked simply, 'What forces are acting on you when you sit on a stool?'. This was an invitation to describe balanced forces in a static system, but the majority of responses which referred to relevant forces mentioned only gravity or weight. This was the case both before and after intervention. Table 6.16 summarises references to gravity and makes clear how extensive is the appreciation of gravity as a force acting on a sitting person, especially post-intervention. The contrast with the low incidence of references to gravity in dynamic situations of other than downward movement the situation in which the ball is moving upwards and the can moving forwards for example - is very apparent.

When you sit on a stool your mass is pulled downward by gravity.

Y9 B H

Table 6.16   Gravity acting on seated child

	Pre-Intervention				Post-Intervention			
	KS1 n.a.	LKS2 n=29	UKS2 n=29	KS3 n=18	KS1 n.a.	LKS2 n=29	UKS2 n=29	KS3 n=18
gravity is acting		17 (5)	41 (12)	67 (12)		45 (13)	59 (17)	89 (16)
weight is acting		7 (2)	7 (2)	22 (4)		3 (3)	3 (1)	11 (2)
gravity/weight not mentioned		75 (22)	52 (15)	11 (2)		45 (13)	38 (11)	-

Another opportunity to examine assumptions about the gravitational force of the Earth arose in the question which showed a school bus moving along the road and asked pupils to, 'Draw arrows on the picture to show the forces acting on the bus.', and 'Put labels on the arrows to show what the forces are.' (The various forces which were identified and some indication of their interactions are described more fully in Section 6.3. The present discussion is limited to a consideration of gravitational force.) The data suggest an age-related increase in the number of children representing gravity as a force in this dynamic situation (31 per cent lower KS2, 59 per cent upper KS2 and 83 per cent KS3). It is of some interest that a sizeable proportion of children at all Key Stages failed to mention gravity or weight in this context (62 per cent lower KS2, 41 per cent upper KS2, 11 per cent KS3).

Figure 6.6

It stops the bus from floating.

Y9 G H

A further opportunity to examine children's appreciation of gravity was available within children's causal explanations of the movement of a floating helium-filled balloon. Children were invited to attach paper clips to the balloon and asked to explain why the balloon moved in the way it did, i.e. horizontally rather than in an upward direction. (The forces on the balloon had been balanced.) At pre-intervention the concept probe was posed to all children in the sample. Children across the three Key Stages tended to focus on the paper clips as preventing the balloon moving upward. Few responded in terms of gravity. Following intervention, it was judged to be inappropriate to probe the very young children's understanding of balanced forces so the concept probe was posed to children at Key Stages 2 and 3 only. Across Key Stage 2 and 3 there was an increase in the number of children using gravity correctly in their explanations. Over one third of the children in Upper Key Stage 2 and Key Stage 3 identified gravity as one of two forces operating on the balloon. At Key Stage Three, one quarter of the children reasoned that gravity was one of a pair of balanced forces.

Table 6.17   Gravity mentioned in connection with helium balloon.

	Pre-Intervention				Post-Intervention			
	KS1 n = 42.	LKS2 n=29	UKS2 n=29	KS3 n=18	KS1 n.a.	LKS2 n=29	UKS2 n=29	KS3 n=18
Gravity mentioned as one of a pair of balanced forces.						3 (1)	3 (1)	28 (5)
gravity mentioned correctly as one of two forces. Not necessarily correct forces, not necessarily balanced	-	-	3 (1)	11 (2)		7 (2)	31 (9)	39 (7)
Gravity mentioned by itself.	-	-	-	11 (2)		-	10 (3)	-
Gravity used incorrectly.	-	-	3 (1)	6 (1)		3 (1)	3 (1)	-
No mention of gravity.	100 (42)	100 (29)	93 (27)	72 (13)		86 (25)	51 (15)	33 (6)

6.2.2   Ideas about friction

The same caveats apply to the following discussion of friction as to the previous consideration of pupils' ideas about gravitational force. Friction needs to be considered by pupils in the context of the system in which movement between two surfaces occurs. For the sake of simplicity and clarity, aspects of beliefs about friction are considered more or less in isolation in this section. The understanding of friction as a force interacting with other forces is considered in Section 6.3.

The first question which precisely targeted the concept of friction was set in the context of riding a bicycle, a pastime assumed to be familiar to all children both in general and in the specifics of how it feels to pedal across different surfaces. It was explained that when a boy rode across a field (contrasted with riding across the playground) he had to pedal harder. 'Why is it harder to ride across grass?', was the question posed to the Key Stage 2 and Key Stage 3 groups, post-intervention only. Table 6.18 summarises the outcomes.

Table 6.18   Difficulty of riding bicycle on grass

	Post-Intervention			
	KS1 n.a.	LKS2 n=29	UKS2 n=29	KS3 n=18
more friction on grass		41 (12)	48 (14)	72 (13)
otherwise expressed		-	7 (2)	-
playground smooth		-	3 (1)	11 (2)
grass bumpy		24 (7)	17 (5)	6 (1)
more gravity on grass		7 (2)	7 (2)	-
less friction on grass		-	-	11 (2)
other responses		28 (8)	17 (5)	-

Familiarity with the idea and the word 'friction' was surprisingly extensive: about two fifths at Lower Key Stage 2, half the pupils at Upper Key Stage 2 and almost three quarters at Key Stage 3 used the term in their explanations. Only two Key Stage 2 pupils expressed the idea of an opposing force on grass with an apparent lack of access to the technical vocabulary.

Ch   There is more grip on the grass.

Int   How does that work?

Ch   It interferes with the tyres, wheels.

Y5 B L

However, amongst the younger children about one quarter were more drawn to the uneven nature of the grassy surface than to the abstraction of 'friction'.

An interesting idea to ponder is the suggestion from some children that there is *more* gravity when riding on grass. Gravity seems to be treated in such instances as a force which opposes movement.

The gravity on the grass tries to pull the bike back.

Y3 B H

Because the gravity is pushing harder. The gravity is acting on his head and his bike, on the handle bars and on the wheels.

Y3 B M

More puzzling is the suggestion by two Key Stage 3 pupils that there was *less* friction associated with riding over the grass. Perhaps for those who think that friction enables or causes movement, greater difficulty in movement results from less friction.

Ch   *Because the playground is smooth and the grass is rough and there is less friction on the grass.*

Int   *What makes you think there is less friction?*

Ch   *Grass isn't very hard. I don't know I've just found it hard to pedal when I've done it.*

Y7 G L

Friction was also directly addressed by a question which presented the situation of a book on a plank in three orientations (see Figure 6.7). Firstly the plank was shown as level and the book still; secondly, the plank was raised a little, the book remaining still; thirdly, the plank was raised higher and the book was shown sliding. Pupils were instructed, 'Write under each drawing if you think that the force of friction is acting on the book.' Table 6.19 presents a summary of responses.

Figure 6.7

Table 6.19   Frictional force between book and plank - summary of responses

	Pre-Intervention				Post-Intervention		
	KS1 n.a.	LKS2 n=29	UKS2 n=29	KS3 n=18	KS1 n.a.	LKS2 n=29	UKS2 n=29	KS3 n=18
1. no, 2. yes, 3. yes - friction opposes movement		24 (7)	21 (6)	44 (8)		52 (15)	28 (8)	56 (10)
1. no, 2. no, 3. yes - no friction without movement		24 (7)	28 (8)	22 (4)		28 (8)	31 (9)	17 (3)
1. yes, 2. yes, 3. no - no friction with movement		7 (2)	3 (1)	6 (1)		7 (2)	10 (3)	11 (2)
1. yes, 2. yes, 3. yes - always friction between surfaces		3 (1)	-	11 (2)		3 (1)	21 (6)	11 (2)
other		41 (12)	48 (14)	17 (3)		10 (3)	10 (3)	6 (1)

The scientifically accurate response, one which recognised that friction is a force which opposes movement between surfaces, was the majority response in the post-intervention interviews. About half the sample at Lower Key Stage 2 and Key Stage 3 suggested that no friction would be acting in the first case (book stationary, plank horizontal), but that friction would be operating when the plank was raised in both cases. The substantial increase in the success rate of the lower Key Stage 2 pupils is worthy of further investigation.

1.   Friction is not acting.

*2   Friction is acting. Two sides are rubbing together and that creates friction.
If friction is greater than the gravity force pulling downwards it stays still.
If gravity force is greater it moves.*

3.   Friction is acting.

Y9 B M

The second group of responses is consistent with the idea that there is no friction without movement, a belief that was fairly widespread but which decreased in incidence quite sharply in the Key Stage 3 group.

1.   No. It is not moving.

2.   No it is not moving.

3.   Yes the book was moving, friction is things moving.

Y4 G L

Many children who indicated no friction associated with the sliding book failed to give any explanation. This reasoning may be influenced by consideration of friction as a force which *impedes* movement. Within this view the instance of a sliding book might suggest no friction. Others asserted that friction *caused* the movement of the book.

Int   What does friction do?

Ch   Friction makes the book go very fast.

Int   How does it do that?

Ch   Friction helps things move.

Y5 B L

We have to be very careful in interpreting these responses. In the case of a wheeled vehicle, it would be correct to assert that 'the road pushes the car'. The force exerted by the road as the wheel rotates against it would not be possible without friction. It would be like attempting to drive on ice. This understanding may have confounded pupils' understanding of friction in the context of the sliding book.

Some Key Stage 2 and Key Stage 3 pupils expressed awareness that friction would be acting in the case of the sliding book and also knew that this force would be less than in the example of the stationary book.

The fourth group summarised in Table 6.19 offered judgements consistent with the view that there is always some friction between surfaces in contact. This view was most likely to have been made explicit in the case of the stationary book on the horizontal plank.

1.   *There is friction because if there was no friction it would be sliding. Friction is in between the table and the book. It keeps it steady.*

2.   *Yes, friction has to stop it moving.*

3.   *Yes, friction is trying to stop it moving.*

                                                                                    Y6 B M

Ch   *There is, but it won't be able to move because there is no slope, friction is all along the plank of wood.*

Int   *Is it there when there is no book?*

Ch   *Yes it's there when there is no book.*

                                                                                    Y6 G M

The question which asked, 'What forces are acting when you sit on a stool?' also elicited some ideas about friction. In this context, interpretation of responses is complicated by the fact that the forces operating on a seated person might be thought of in more than one way. For example, if a person is seated in a stationary position, the forces acting on the body are gravitational force and the force exerted in return by the stool. Alternatively, the person may be thought of as sitting in a less stable position or even as twisting or otherwise moving, in which case friction most certainly will oppose the movement between the body and the surface of the stool. Because of these two possible viewpoints, the references to friction in Table 6.20 distinguish between those which offer justification for the force of friction and those which mention it without such justification.

There were no justifications amongst the pre-intervention responses, but these did occur post-intervention.

Ch   *Friction between you and the stool .*

Int   *How does it work?*

Ch   *It keeps you steady.*

Int   *Is there friction anywhere else?*

Ch   *Friction between the stool and the floor.*

Int   *How does that work?*

Ch   *It stops you sliding.*

Y7 B H

More commonly obtained from the younger pupils in the sample were suggestions that friction would be operating without any correct justification.

The chair is a surface and you're sitting on it and it's stopping you from falling. Friction is straight up.

Y3 G M

Table 6.20   **Friction acting on seated child.**

	Pre-Intervention				Post-Intervention			
	KS1 n.a.	LKS2 n=29	UKS2 n=29	KS3 n=18	KS1 n.a.	LKS2 n=29	UKS2 n=29	KS3 n=18
friction with explanation	-	-	-	-	-		3 (1)	22 (4)
friction without explanation			7 (2)	28 (5)		10 (3)	21 (6)	22 (4)
friction not mentioned		100 (29)	93 (27)	72 (13)		90 (26)	75 (22)	56 (10)

Responses referring to friction will be considered next by returning to the question about the forces acting on the bus (discussed in relation to gravity in section 6.2.1 above and in the context of interacting forces in Section 6.3.2.5 below). Firstly, arrows drawn to represent friction acting in association with the body of the bus will be considered. The main feature of this form of response was its sporadic occurrence. Only five pupils (six per cent of the combined KS2 and KS3 sample to whom this question was posed) drew arrows labelled 'friction' in their pre-intervention responses, and five post-intervention. Remarkably, there were no pupils in common between the two sets of five, which seems to confirm the elusiveness of appreciation of frictional force in this form.

An arrow drawn to represent friction associated with the wheels was slightly more common than those associated with the body of the bus, (six per cent of KS2/3 pupils pre-intervention, compared with 50 per cent post-intervention). In situations such as the book on the plank the force of friction acts in the opposite direction to any movement. However, for wheeled vehicles, the frictional force opposes the movement of the *tyres* in contact with the road and is, therefore in the *same* direction as the movement of the vehicle. No pupil offered a forward pointing arrow unequivocally associated with the force of friction on a wheel.

A surprisingly small proportion of children drew their arrows pointing backwards, (three per cent Lower KS2, 14 per cent Upper KS2 and 17 per cent KS3), i.e in the direction from which the bus has travelled.

Figure 6.8

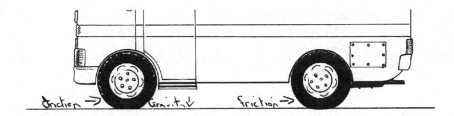

Y6 B H

Many more pupils (seven per cent Lower KS2, 28 per cent Upper KS2 and 67 per cent KS3) drew their arrows pointing in some other direction than clearly horizontal (forwards or backwards) or vertical.

Figure 6.9

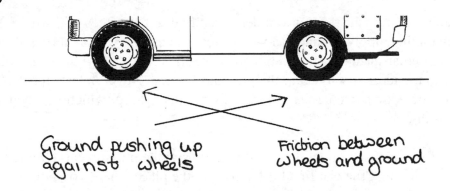

Y9 B M

A small proportion of pupils drew curved arrows. Whereas such a representation violates the fundamental Newtonian notion that forces act in straight lines, in these responses, it more likely indicated a lack of understanding of the conventional use of arrows to represent forces.

Figure 6.10

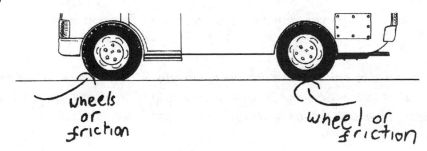

Y3 G H

Some of the more extended verbal responses obtained during interviews enabled pupils to elaborate their ideas beyond what it was possible to illustrate in drawings alone. One idea that emerged was that a frictional force depends on the *speed* at which an object is travelling. When stationary or moving slowly, friction was not believed to be acting (or was thought to be acting only minimally).

Because the wheels are going fast, the wheels start rubbing. It comes from the wheels and the ground. The bus driver has his foot on the pedals. The wheels are going so fast it causes friction. If it goes slowly there is not so much friction. When two things rub together the air begins to cause friction as well.

Y3 G H

In summary, it is apparent that children did not show much evidence of understanding of the frictional forces associated with the turning wheels of a bus. Most of those who drew arrows labelled as representing friction presented them at various angles around the wheels, some even curved. This lack of conceptual understanding was further confounded by the lack of understanding of the conventions associated with representing forces with arrows.

6.2.3   *Air resistance*

The *caveats* about treating air resistance independently of other forces with which inevitably, it would be interacting, apply as much as was the case in the discussion above of gravity and friction. As a particular instance of a frictional force, this discussion of air resistance must also be related to the ideas discussed in the previous section. Nevertheless, for the sake of clarity of focus, there are some issues associated with this concept which will benefit from a separate discussion.

In the context of the question about the can dropped from the moving car, children were asked, 'Why do you think the can first hit the road at the place where you put the letter c?'. This question posed a complex problem because of the changing co-ordinates of the relative positions of car, can and road. Nonetheless, what was presented was a situation which most children could readily imagine as real.

In their pre-intervention responses, no children used 'air resistance' as a scientist would, to describe a force that would slow the forward movement of the can. Those children who mentioned air resistance at all, described it as a force which would move the can backwards.

Because its going fast the can will blow back. Air resistance forces it back.

Y6 G H

The pressure is against the car because the car is moving. Air resistance, when something is moving the can will always have some resistance in the opposite direction and when the can is dropped it doesn't have any engine of its own so the can is forced back.

Y6 B H

The post-intervention interview responses suggest that reasoning in terms of air resistance was not very stable amongst the Lower Key Stage 2 pupils, while the Upper Key Stage 2 pattern remained more consistent. There were some shifts at Key Stage 3 which, while of small percentages, are interesting in providing insights as to the kinds of movement in understanding that might be achieved. Only four of the 18 Key Stage 3 pupils referred to air resistance. One stayed with the idea of air resistance being responsible for the can moving backwards. The other three showed appreciation that the force and movement of the can would be slowed although one was of the opinion that the air resistance would exactly cancel out the forward movement of the can.

Ch   Because it will drop straight down because it is still moving with the car.

Int   How will that happen?

Ch   When it is released it will start slowing down with the air resistance.

Y7 B H

Ch   When you drop the can it doesn't drop directly down, it drops forward and then air resistance pushes it, for example, acts on it so it goes down straight. The vehicle is moving forward so the can moves forward and because there is no more energy getting to the can, so force take over.

Int   What are these forces?

Ch   Air resistance and gravity take over.

Y9 B L

The can is pushed forward by the speed and pulled back by air resistance resulting in the can dropping roughly below where it was dropped from.

Y9 B M

It had been noticed that many pupils used the expression 'wind' or 'wind resistance' when describing opposition to the can's forward movement through the air. This form of wording was recorded when responses were coded; outcomes are summarised in the lower half of Table 6.21. It can be seen immediately that 'wind' was a far more popular interpretation of the phenomenon than 'air resistance'.

The distinction between 'air resistance' and 'wind' (or 'wind resistance') is not a trivial one in terms of the sense which children impose on this event. When physicists objectify the event of movement across two surfaces in contact, the important consideration is that the movement is in different directions (or at different speeds in the same direction). For example, air resistance between a car and the air is the same whether the car moves through the air (as when it is driven on a road), or the air moves across the car (as in a 'wind tunnel'). The latter is often the more convenient method of testing a car's body or an aeroplane's wing for its aerodynamic characteristics. (Similarly, boat hulls are tested in a flume.) Children clearly take time to achieve understanding of this equivalence, though their use of the word 'streamlined' is not uncommon. A moving body of air is felt because it exerts a force as it passes across and around a person. If the wind is strong enough, children will have felt the need to brace themselves or may even have been caused to stagger. They may have seen umbrellas blown inside out or even more extreme damage caused by the wind: trees uprooted, buildings damaged. In contrast stationary air is intangible. The results summarised in the lower half of Table 6.21 suggest that the ideas of the force of a moving body of air (i.e. what we commonly call 'wind') and the force opposing the movement of a body through a static body of air (i.e. 'air resistance') are frequently confounded in children's thinking.

Table 6.21   Dropping can from moving car-effects associated with movement through air

	Pre-Intervention				Post-Intervention			
	KS1 n=42	LKS2 n=29	UKS2 n=29	KS3 n=18	KS1 n=42	LKS2 n=29	UKS2 n=29	KS3 n=18
References to air resistance								
a.r. slows can	-	-	-	-	-	-	-	11 (2)
a.r. cancels forward momentum	-	-	-	-	-	-	-	6 (1)
a.r. moves can backwards	-	17 (5)	7 (2)	11 (2)	-	7 (2)	7 (2)	6 (1)
a.r. not mentioned	100 (42)	83 (24)	93 (27)	89 (16)	100 (42)	93 (27)	93 (27)	78 (14)
References to wind								
wind blows can forwards	7 (2)	7 (2)	7 (2)	-	-	7 (2)	3 (1)	6 (1)
wind blows can backwards	17 (5)	28 (8)	55 (16)	39 (7)	19 (8)	66 (19)	69 (20)	39 (7)
wind/wind resistance not mentioned	83 (35)	66 (19)	38 (11)	61 (11)	81 (34)	28 (8)	28 (8)	56 (10)

Because the movement through the air would be sensed against the skin in this context, it is interpreted as wind. The frequencies in Table 6.21 show that most children describe this 'wind' as blowing the can *backwards*.

Ch   *If the car's going fast the can is going to be pushed back. The force will push it back.*

Int   *What force?*

Ch   *The wind would be pushing the car and it would come off the wind shield down the side of the car and it would push the can. The car is going through the wind and pushing it either side of the car.*

Y9 G L

Ch   *Because the force acting on the wind is pulling it back. If it's going fast it will go right back because the wind gets stronger and pushes it.*

Y6 G H

About two thirds of the Key Stage 2 responses were of this kind, compared with one third at Key Stage 3. This was an increase compared with the Key Stage 2 responses of the same kind pre-intervention, while the Key Stage 3 proportion of this form of response remained static.

A small number of children described 'wind' as blowing the can 'forwards'.

Ch *It will drop to the front because they are going fast so the can will drop. They are going fast and the push force of the wind will push it to the front.*

Int *Where does this wind come from?*

Ch *It comes from the back of the car. It's called 'push thrust'.*

Y5 G H

Ch *Because the wind pushes it forward.*

Y7 B L

Because of the shifting co-ordinates between the forward-moving car, the falling can and the road, there is the potential for ambiguity over the net directions which children understand and intend to communicate. For example, when the can was described as moving 'backwards', interviewers probed to ensure that what was intended was more than the forward movement being slowed.

The question which asked children to, 'Draw arrows on the picture to show the forces acting on the bus.' had the potential to elicit similar ideas to those described in relation to the can dropped from the moving car.

Figure 6.11   **Figure 6.12**

wind pushing against front

Y9 G M

slows it down wind that comes when the bus is moving

Y5 B M

References to 'wind' or 'wind resistance' were almost totally absent in the post-intervention responses, with just a single Lower Key Stage 2 pupil offering the idea that 'wind resistance' operated at the front of the bus to oppose its forward movement. The incidence of 'air resistance' responses, compared with those for the dropped can, was greatly increased, all associated with an arrow pointing towards the front of the bus, opposing it's forward movement. This form of response was offered by three per cent of pupils at Lower Key Stage 2, ten per cent at Upper Key Stage 2 and 56 per cent at Key Stage 3.

Care was taken to code 'wind resistance' and 'wind' responses separately, and the latter are summarised at the bottom of Table 6.22. Twenty seven per cent of lower Key Stage 2, twenty per cent of Upper Key Stage 2 and eleven per cent of Key Stage 3 drew arrows labelled 'wind'. A greater proportion of these arrows were drawn at the front of the bus, (opposing its forward movement) post-intervention.

The differences in responses when the can and the bus are the subject of consideration may be attributable to the very different sizes and masses involved, as well as the fact that the bus has its own mode of propulsion while the can does not.

Table 6.22   Forces on moving vehicle - effects associated with movement through air

	Post-Intervention			
	KS1 n.a.	LKS2 n=29	UKS2 n=29	KS3 n=18
Reference to 'air resistance'				
a.r arrow, front, opposing		3 (1)	10 (3)	56 (10)
a.r. arrow, front, assisting		-	-	-
a.r. arrow, rear, assisting		- -	- -	- -
no a.r. arrow		97 (28)	90 (26)	44 (8)
Reference to 'wind resistance'				
w.r. arrow, front, opposing		3 (1)	-	-
w.r. arrow, rear, assisting		-	-	-
no w.r. arrow		97 (28)	100 (29)	100 (18)
Reference to 'wind'				
wind arrow, front, opposing		14 (4)	17 (5)	11 (2)
wind arrow, side, opposing		3 (1)	-	-
wind arrow, front, assisting		-	-	-
wind arrow, rear, assisting		10 (3)	3 (1)	-
no wind arrow		72 (21)	79 (23)	89 (16)

6.3   *Balanced and Unbalanced Forces*

This section explores the extent to which children apply their knowledge and understanding of forces and the associated scientific language to a number of everyday events and experiences. In so doing it considers their appreciation of the fact that, in these events and experiences, objects are acted upon by several forces simultaneously. It sets, therefore, their understanding of specific forces, considered in Sections 1 and 2, into more complex situations.

For the purposes of initial discussion of the data the section has been sub-divided into:

6.3.1 Balanced forces acting on stationary objects

6.3.2 Unbalanced forces acting on moving objects

6.3.3 Balanced forces acting on moving objects

6.3.1   *Balanced forces acting on stationary objects*

6.3.1.1   *Size of the reaction force*

Previous research (Erikson and Hobbs, 1978; Minstrell, 1982) has indicated that the concept of reaction force presents children with considerable difficulty. The assumption underlying the particular probe discussed here (which made use of a top-pan balance) was that children would be able to appreciate a reaction force most readily if they were to receive a direct tactile experience of one. In consequence, at the pre-intervention stage the children were provided with a top-pan balance, asked to press down on it to give a reading of 10N, then to say whether the pan was exerting a force on the pressing hand. If so, they were asked to predict the size of this force. About 40 per cent of the children at Key Stage 2 recognised that there was a force on the hand from the pan but of these, only a half were able to predict its size accurately. All but one of the Key Stage 3 children accepted that a force was acting and three quarters of them were able to indicate an understanding that its size was equal to that of the push from the hand.

The sizeable proportion of Key Stage 2 children (90 per cent Lower KS2 and 60 per cent Upper KS2) *not* able to confirm the perception of a reaction force of equal magnitude in these most favourable circumstances emphasises the difficulty experienced with this concept. The extent of the difficulty does appear, in this context, to be age-related, as the details of Table 6.23 show. However, the data from the seated child probe (discussed next) are much less clear in this regard.

Following the intervention activities the probe was repeated at Key Stage 3 only, but without the direct experience of pushing on a balance. No improvement in the number accepting the equality of the reaction force was recorded.

This probe highlights the difficulties many children experience in conceptualising the idea of reaction force even when presented in its most tangible, perceptible manifestation. Amongst those able to confirm the existence of reaction force as a phenomenon, the probe reveals that most are not able to recognise that a static situation requires the forces acting to be in *balance*.

Table 6.23   Size of reaction force

	Pre-Intervention				Post-Intervention			
	KS1 n.a.	LKS2 n=29	UKS2 n=29	KS3 n=18	KS1 n.a.	LKS2 n.a.	UKS2 n.a.	KS3 n=18
reaction force equal		10 (3)	38 (11)	72 (13)				67 (12)
reaction force less		17 (5)	3 (1)	11 (2)				17 (3)
reaction force more		7 (2)	7 (2)	11 (2)				11 (2)
no reaction force		34 (10)	3 (1)	-				-
no response		31 (9)	48 (14)	6 (1)				6 (1)

6.3.1.2   Forces acting on a seated child

Children's experience of sitting on a chair was used to probe their understanding of the forces acting in this static situation. It was anticipated that, particularly for the younger children, the whole-body experience would enhance a more correct interpretation of the downward force on the body due to gravity being balanced by the upward reaction force of the chair.

Table 6.24 shows the responses to the question 'What forces are acting on you when you sit on a chair?'. Gravity or weight was the most commonly mentioned force. However, prior to the intervention a significant number failed to include it. This omission was clearly age-related, being 75 per cent at lower Key Stage 2 falling to 11 per cent at Key Stage 3. This perhaps reflects the view that gravity ceases to act when objects stop moving downwards. The intervention activities resulted in a substantial reduction in these figures to 45 per cent and zero respectively.

At the pre-intervention stage only one child (KS3) indicated that there is a force from the stool. Possibly sitting is experienced too frequently for it to invoke the idea of a reaction force. The intervention activities, involving quantification and force diagrams, appear to have induced a recognition of such a force in about one quarter of all of the children. Surprisingly, this improvement in understanding did not increase with age.

On the chair: the forces push down to keep the chair steady

On you: The force keeping you up

Y7 B M

Table 6.24   Forces acting on seated child

	Pre-Intervention				Post-Intervention			
	KS1 n.a.	LKS2 n=29	UKS2 n=29	KS3 n=18	KS1 n.a.	LKS2 n=29	UKS2 n=29	KS3 n=18
Gravity								
gravity is acting		17 (5)	41 (12)	67 (12)		45 (13)	59 (17)	70 (16)
weight is acting		7 (2)	7 (2)	23 (4)		10 (3)	3 (1)	11 (2)
gravity/weight not mentioned		75 (22)	52 (15)	11 (2)		45 (13)	38 (11)	-
Reaction Force								
'reaction force' mentioned		-	-	-		-	3 (1)	-
reaction force otherwise expressed		-	-	6 (1)		24 (7)	24 (7)	22 (4)
reaction force not mentioned		100 (29)	100 (29)	94 (17)		76 (22)	72 (21)	78 (14)
Balanced Forces								
two correct forces, balance mentioned		-	-	-		3 (1)	7 (2)	-
incorrect forces, balance mentioned		3 (1)	-	6 (1)		10 (3)	10 (3)	6 (1)
two correct forces, balance not mentioned		-	-	6 (1)		7 (2)	17 (5)	22 (4)
only weight/gravity, balance not mentioned		28 (8)	48 (14)	83 (15)		41 (12)	31 (9)	72 (13)
only reaction force, balance not mentioned		-	-	-		3 (1)	-	-
no correct forces, balance not mentioned		69 (20)	52 (15)	6 (1)		34 (10)	34 (10)	-

A comparison of the data from this probe with those discussed earlier in relation to the top-pan balance suggests that not all of the children who were able to predict the size of a reaction force once such a force had been pointed out to them would, in the absence of prompting, have considered the existence of such a force in the context of sitting on a stool. Nevertheless, the implication from the data is that with appropriate intervention the concept of reaction force can be made accessible to children.

For the large proportion of children considering that only a single force is acting in this situation there can be no question of balance.

110

Gravity - No more forces. Gravity pushes you down so you don't float off.

Y7 G H

When you sit on a stool your mass is pulled down by gravity. The stool just keeps you up.

Y9 B H

Even those who responded with both correct forces, in most cases did not mention the balance between them. It may be that a more productive teaching sequence would place an understanding of the need for *balanced* forces in static situations somewhat earlier than is common in order to use the logic of this idea to dictate a need for *reaction* forces.

6.3.1.3   *Forces acting on a helium-filled balloon*

Children were provided with a helium-filled balloon with a string attached. They were then challenged to add to the string small objects (e.g. paper clips) in just sufficient numbers to prevent the balloon from moving either upwards or downwards. When they had succeeded in this task, the question was posed, 'What can you say about the forces acting on the balloon when it is like this?'.

Figure 6.13

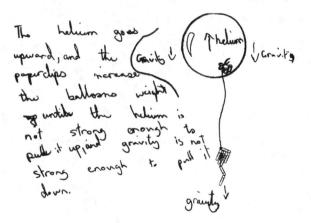

Y6 B H

The helium wants to go up and it can do that until it hits the ceiling. But when you have paperclips, the weight pulls it down slowly, or it will hang in mid air if you hang the right amount.

Fig 6.14

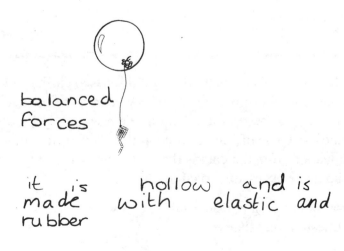

Y6 G M

Table 6.25   Forces acting on helium-filled balloon

	Pre-Intervention				Post-Intervention		
	KS1 n=42	LKS2 n=29	UKS2 n=29	KS3 n=18	KS1 n.a.	LKS2 n=29	UKS2 n=29	KS3 n=18
Two Forces								
two correct forces/objects, balance mentioned	-	10 (3)	17 (5)	22 (4)		14 (4)	24 (7)	34 (6)
two incorrect forces, balance mentioned	-	-	3 (1)	6 (1)		21 (6)	7 (2)	50 (9)
two correct forces/objects, balance mentioned	3 (1)	21 (6)	24 (7)	22 (4)		17 (5)	28 (8)	6 (1)
Single Force								
gravity/weight only acting	3 (1)	21 (6)	31 (9)	39 (7)		-	10 (3)	6 (1)
paper clips hold it down	55 (23)	24 (7)	14 (4)	6 (1)		24 (7)	14 (4)	6 (1)
air supports balloon	5 (2)	10 (3)	-	-		3 (1)	7 (2)	-
helium holds it up	-	-	3 (1)			7 (2)	-	-
Other Explanations								
no forces acting	-	-	-	-		3 (1)	-	-
nature of balloon	-	-	3 (1)	-		3 (1)	-	-
other responses	31 (13)	3 (1)	3 (1)	6 (1)		3 (1)	3 (1)	
don't know	-	-	-	-		3 (1)	7 (2)	-
no response	5 (2)	10 (3)	-	-		-	-	-

The data in Table 6.25 suggest that being physically involved in the act of countering the upward force of the balloon by adding objects in just the right quantity resulted in a significantly greater proportion of children with the notion of a balance in this static situation, even at the pre-intervention stage, as compared with the top-pan balance and stool situations. Post-intervention, many of those who were unable correctly to identify the forces involved were nevertheless aware of the necessity of balance.

However, this concentration on the added objects proved so powerfully attractive to many that they excluded all other forces from their interpretations both pre-and post-intervention. Perhaps some of these respondents felt that it is simply the nature of helium-filled balloons to rise. This attribute seems not to be considered to be in need of a name. On the other hand, a force to counteract this tendency of the balloon to rise is more familiar.

112

An understanding of the helium balloon behaving in the way it did as the result of its interaction with its environment - the air in which it floated - is an abstract idea. The helium and
the air are both invisible. It was unlikely that children would draw analogies with objects
floating in water without specific support from their teachers. Upthrust, as a phenomenon
common to all fluids, was understandably not well understood by these children and this
lack of awareness led to many describing the upward force on the balloon as the 'balloon
force' or the 'force of the helium', etc. Such responses were accepted as legitimate at this
level. A few, however, referred to the upward force as 'air resistance'.

6.3.2 Unbalanced forces acting on objects

6.3.2.1 Forces on a bottle using the arrow convention

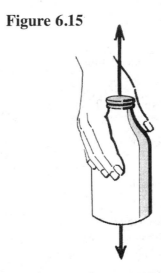

Figure 6.15

The children were asked to interpret a line diagram
of a bottle showing the forces acting upon it (gravity
and a lift from a hand) represented by arrows. The
arrows were in accordance with the convention
regarding size and direction. Section 6.1.7 discussed
the children's responses in terms of their appreciation of the direction and magnitude of the forces
represented by arrows.

In addition, however, it is possible to use their
responses as indications of their understanding of
the probable *nature* of the forces acting and the
resulting *movement* of the bottle.

The fact, noted previously, that children frequently omit to mention gravity or weight as a
force, is exemplified again in the responses summarised in Table 6.27. However, the intervention activities here effected a considerable change at Key Stage 3 - from 39 per cent to
89 per cent. For the younger children, the most plausible interpretation of the data is that
they were unfamiliar with the arrow convention.

This unfamiliarity almost certainly also accounts for the failure of the majority of younger
children to name the upward force on the bottle. It appears to have been less of a problem
for those at Key Stage 3 as correct responses were given by 44 per cent at pre-intervention
and 67 per cent following intervention.

Table 6.26 summarises data elicited regarding children's ideas as to what the two arrows represent in the sense of what quality of force they assumed to be acting.

Table 6.26   Arrow representation of forces

	Pre-Intervention				Post-Intervention			
	KS1 n.a.	LKS2 n=29	UKS2 n=29	KS3 n=18	KS1 n.a.	LKS2 n.a.	UKS2 n.a.	KS3 n=18
Nature of downward force								
gravity/weight		17 (5)	28 (8)	39 (7)				89 (16)
air resistance		-	3 (1)	-				-
push from hand		-	-	11 (2)				-
force not named		83 (24)	69 (20)	50 (9)				11 (2)
Nature of upward force								
pull from hand		14 (4)	7 (2)	44 (8)				67 (12)
friction		-	-	6 (1)				6 (1)
air resistance		7 (2)	3 (1)	-				6 (1)
gravity/weight		-	-	6 (1)				6 (1)
push from liquid		-	7 (2)	-				-
force not named		79 (23)	83 (24)	44 (8)				17 (3)

Table 6.27   Arrow representation of forces - resultant movement.

	Pre-Intervention				Post-Intervention			
	KS1 n.a.	LKS2 n=29	UKS2 n=29	KS3 n=18	KS1 n.a.	LKS2 n.a.	UKS2 n.a.	KS3 n=18
bottle moves upwards		7 (2)	7 (2)	28 (5)				44 (8)
bottle moves downwards		-	-	-				6 (1)
bottle does not move		-	3 (1)	-				-
bottle moves up and down		3 (1)	3 (1)	6 (1)				6 (1)
movement not mentioned		90 (26)	86 (25)	67 (12)				44 (8)

114

The children's lack of knowledge of the convention also prevents the data concerning the movement of the bottle from being reliably informative for those at Key Stage 2. The intervention activities for Key Stage 3 children were able to effect a significant increase in correct movement responses (i.e. responses consistent with the fact of the longer upward arrow) from 28 per cent to 44 per cent. (See Table 6.27).

The down arrow is the force of gravity and the up arrow is the force of the persons hand on the bottle. The upward arrow is longer so the upward force is greater than the downward force so the bottle is being picked up or lifted.

Y9 B H

If knowledge of the convention is assumed for the remainder (which is by no means a certainty), then the fact that unbalanced forces cause changes to movement was not well understood.

It was noted by teachers that once the arrow convention of representing the direction and magnitude of the forces is understood by children, it can be used as an extremely helpful assessment tool. Indeed, the arrow convention is an excellent example of Representational Redescription: movement in space can be represented graphically, in two dimensions, as an explicit check on interpretations of the outcomes of various forces acting.

6.3.2.2   Starting to move on a bicycle

Children were asked, 'What do you have to do to make your bicycle start moving?'. This was one of the earliest questions posed. The thinking behind it was to offer an open opportunity to children to describe a familiar, whole body activity in terms of motion and forces, or other vernacular descriptions which they might favour. Prior to intervention, less than half of the children at Key Stages 1 and 2 used 'push' or 'force' in their responses. The most common descriptions referred to movements of their legs or parts of the bicycle. On the other hand, nearly three quarters of the Key Stage 3 children provided responses which recognised that a force is needed to initiate movement.

The recommended intervention activities provided opportunities and encouragement for children to describe everyday activities using more specifically force-related words. The evidence gathered during the pre-intervention interviews gave strong indications that younger children, in particular, were experiencing difficulties with the word 'force'.

Table 6.28   Starting to move on bicycle

	Pre-Intervention				Post-Intervention			
	KS1 n=42	LKS2 n=29	UKS2 n=29	KS3 n=18	KS1 n=42	LKS2 n.a.	UKS2 n.a.	KS3 n.a.
use force	-	-	14 (4)	-	-			
push pedal/ground	14 (6)	48 (14)	28 (8)	72 (13)	71 (30)			
ride/move legs	57 (24)	45 (13)	38 (11)	28 (5)	24 (10)			
wheel/chain turns	12 (5)	-	14 (4)	-	-			
other responses	14 (6)	7 (2)	7 (2)	-	5 (2)			
no response	3 (1)	-	-	-	-			

Subsequent discussion led to an agreement that Key Stage 1 children should be encouraged to continue to use words such as 'push' and 'pull' in their explanations with no expectation of an early introduction of 'force'. The repeat of the bicycle question at post-intervention for Key Stage 1 children revealed a considerable increase in responses using 'push', from 14 per cent to 71 per cent. (See Table 6.28). These children were thinking of the activity in a manner which revealed a closer focus on the *causes* of movement, using an age-appropriate vocabulary.

What do you have to do to make your bicycle start moving?

You, Push The paddels with your feet.

Y2 G M

What do you have to do to make your bicycle start moving?

you puch the pedle and the chan mov and the well go

You push the pedal and the chain moves
and the wheel goes

Y2 G M

The above example illustrates the kind of observation-related reasoning which even five to six year old pupils can be encouraged to use. Such complete sequences of causal reasoning are by no means commonplace in this age group.

6.3.2.3 Forces on a can dropped from a moving car.

The children were presented with a line drawing of a car which they were told was travelling 'forwards quite fast'. The children were invited to mark on the drawing the position of first impact on the road of a can dropped from the car and to give their reasoning. Their responses were categorised as 'ahead of', 'directly beneath' or 'behind' the point of release and are summarised in Table 6.29.

Table 6.29   Dropping can from moving car - position of impact on road.

	Pre-Intervention				Post-Intervention			
	KS1 n=42	LKS2 n=29	UKS2 n=29	KS3 n=18	KS1 n=42	LKS2 n=29	UKS2 n=29	KS3 n=18
ahead	21 (9)	10 (3)	14 (4)	17 (3)	17 (7)	7 (2)	3 (1)	33 (6)
directly beneath	55 (23)	28 (8)	7 (2)	17 (3)	67 (28)	24 (7)	3 (1)	11 (2)
behind	21 (9)	62 (18)	79 (23)	67 (12)	17 (7)	69 (20)	93 (27)	56 (10)
no response	3 (1)	-	-	-	-	-	-	-

Intervention for the Key Stage 3 sample doubled the number of children with correct impact predictions, taking the proportion from one sixth to one third. Although the data in Table 6.30 indicate that approximately one fifth of Key Stage 1 children at both pre-and post-intervention stages correctly predicted an 'ahead' impact, closer scrutiny of their responses showed that their reasoning did not involve consideration of the forward momentum of the can. For the majority of these children, the can went where it rolled or was thrown (see Table 6.32). The very small proportion of Key Stage 2 children who at pre-intervention had 'ahead' predictions was reduced yet further (from 12 per cent to five per cent) by the readjustments in their thinking brought about by the intervention activities.

Prior to intervention, a little more than a half of Key Stage 1 children indicated that the can would land directly beneath its point of release. This fraction was increased to two thirds post-intervention. It appears that the drop was considered by them in isolation from the other influential factors.

The overwhelmingly popular response, at both pre- and post-intervention stages, was that the can would land in the 'behind' position. The reasons given in the vast majority of cases were based upon a misinterpretation of the direct experience of air hitting a hand or face protruding from a moving vehicle. This interpretation has air moving backwards past the car rather than the car passing through (relatively) static air. The logical conclusion of thinking based on this misinterpretation is that the 'wind' will blow the can backwards. This form of reasoning was discussed previously in Section 6.2.3.

Although in other situations pupil's seemed intuitively to use ideas in a manner which closely approximates the scientific idea of momentum, in this instance, the idea of 'wind' or 'air resistance' seemed to overwhelm any other consideration. Momentum responses are summarised in Table 6.30.

Table 6.30   Dropping can from moving car - can has momentum

	Pre-Intervention				Post-Intervention			
	KS1 n=42	LKS2 n=29	UKS2 n=29	KS3 n=18	KS1 n=42	LKS2 n=29	UKS2 n=29	KS3 n=18
can has forward momentum	-	-	-	-	-	-	-	11 (2)
can continues to move forward	-	-	3 (1)	17 (3)	-	-	-	28 (5)
car is moving fast	-	-	3 (1)	-	2 (1)	3 (1)	3 (1)	-
no mention of momentum idea	100 (42)	100 (29)	93 (27)	83 (15)	98 (41)	97 (28)	97 (28)	61 (11)

Even after intervention, very few at the lower Key Stages and about two fifths at Key Stage 3 were able to base their thinking on the fact that the forward movement of the can prior to release would continue after it. This is another example of children's tendency to think of forces coming into play only when threshold levels have been reached. In this instance, the threshold might be thought of as being limited or held back by the opposing effect of movement through air. It is likely that the experience of empty drinks cans as very light, low density objects, capable of being blown along a street has strongly influenced chidren's judgements. It may be possible that an introduction of the concept of momentum at an appropriate stage in teaching would assist understanding of phenomena such as dropped 'passively moving' objects.

The dropped can question provided more evidence of mention of gravity being omitted in instances where more than one force needs to be considered. Even after intervention, very few children in the earlier Key Stages and only 22 per cent at Key Stage 3 included gravity in their explanations. Again, it is possible that children think of the effects of gravity being suppressed by other, more powerful, forces. In other words, it could be that gravity has not been overlooked so much as been deemed not to have been brought into play. Children frequently used the metaphor of war, battle or struggle between forces. In this instance, gravity might have been considered to be overwhelmed by more active and powerful forces. If gravity is considered to have been 'beaten' by other forces, it is unlikely that it will be mentioned.

A sizeable minority both pre-and post-intervention provided explanations that did not include consideration of the forces acting. The responses of this kind at Key Stage 1, involving throwing and rolling, were mentioned above. The most common of such responses at the other Key Stages considered the position of the can relative to the car, not the road. In consequence children gave 'behind' answers because they had in mind that the car had moved forwards as the can was falling. Even careful probing at interview could not deflect some children (for example 31 per cent at upper KS2) from this viewpoint.

Ch   *The car is going forwards and the can is going down. So it goes at an angle.*

Int   *Why does it go backwards?*

Ch   *Because the can isn't moving backwards it is going down because the car is going forwards. It goes at an angle.*

Int   *Why does it do that I wonder? You've said - 'The can is going backwards. The car is going forwards.'*

Int   *It (the can) is just going down. Because the car is going forwards it seems as if it is moving backwards.*

Table 6.31   Dropping can from moving car - explanations not mentioning forces.

	Pre-Intervention				Post-Intervention			
	KS1 n=42	LKS2 n=29	UKS2 n=29	KS3 n=18	KS1 n=42	LKS2 n=29	UKS2 n=29	KS3 n=18
can thrown	43 (18)	7 (2)	-	-	36 (15)	-	-	-
can rolls	12 (5)	-		-	5 (2)	7 (2)	-	-
can goes there	12 (5)	7 (2)		-	29 (12)	7 (2)	-	-
car moves forward	5 (2)	24 (7)	34 (10)	22 (4)	7 (3)	21 (6)	31 (9)	11 (2)
can is heavy	3 (1)	7 (2)	3 (1)	-	2 (1)	3 (1)	-	-
no response of this kind	26 (11)	55 (16)	62 (18)	78 (14)	21 (9)	62 (18)	69 (20)	89 (16)

6.3.2.4   *Forces causing a ball to bounce*

The children were asked to explain what makes a ball bounce when it is dropped onto the playground. The important attributes of the content of this situation were deemed (by the research team) to be the familiarity of the experience and the visible and tangible elasticity of the ball. It was anticipated that these aspects would encourage and enable children to envisage and discuss reaction force by reference to the visible compression of the ball on impact with the hard playground surface. In the event, this dynamic situation appeared not to be any more effective than the static 'seated' child in eliciting the idea of a reaction force. Only a handful of pupils at both interview stages gave responses which indicated an awareness of the playground's role in the bouncing reaction of the ball.

Gravity pulls it down but the force applied at the other side is enough to send it back.

Y9 B H

Table 6.32   Bouncing ball - reaction force

	Pre-Intervention				Post-Intervention			
	KS1 n.a.	LKS2 n=29	UKS2 n=29	KS3 n=18	KS1 n=42	LKS2 n=29	UKS2 n=29	KS3 n=18
'reaction force' mentioned	-	-	-	-	-	3 (1)	6 (1)
reaction force otherwise expressed		10 (3)	6 (2)	6 (1)	2 (1)	17 (5)	7 (2)	6 (1)
'bounce' force		3 (1)	-	-	2 (1)	-	-	-
reaction force not mentioned		86 (25)	93 (27)	94 (17)	95 (40)	83 (24)	90 (26)	89 (16)

However, in marked contrast to the older children, 33 per cent of those at Key Stage 1 and 24 per cent at lower Key Stage 2 recognised the need for the playground to be *hard*. This is arguably as a result of the younger children responding from their personal experience that the harder the surface the better the bounce. There may be a similar reason for these same children considering that the 'bounce' is related to the force of the downward throw.

when the ball hits a hard surface it will come up again because of the strong downward force

Y3 B H

At the pre-intervention stage this probe asked Key Stage 2 and 3 children to draw the ball at and just after the point of impact with the playground and then to explain their drawings.

One fifth of the Key Stage 2 children and two thirds at Key Stage 3 demonstrated an understanding that the ball was deformed and then reformed.

i) just before it hits the table	ii) as it hits the table	iii) as it bounces upwards again

Explain why you drew the balls the way you did.
It is because when you drop it, it bends in and then expanded.
You can hardly see the little dint where it squashes up a tiny bit.
When it comes back out it causes it to go back in the air again.

Y6 B H

After intervention children at all Key Stages were asked to explain why a ball bounces. There was no requirement for any drawing. In these circumstances the percentages of those mentioning the changing shape of the ball or the consequential effects on the air pressure inside it fell dramatically. No Key Stage 1 child mentioned the change; less than 10 per cent at Key Stage 2 and only 28 per cent at Key Stage 3 did so. These data (Table 6.33) would seem to suggest that the use of appropriate drawings greatly enhances understanding of this phenomenon.

A side collapses in and pushes back out again, forcing it back up into the air.

Y6 B H

Table 6.33 Bouncing ball - change of shape

	Pre-Intervention				Post-Intervention			
	KS1 n.a.	LKS2 n=29	UKS2 n=29	KS3 n=18	KS1 n=42	LKS2 n=29	UKS2 n=29	KS3 n=18
ball deformed then reformed		17 (5)	24 (7)	61 (11)	-	-	3 (1)	11 (2)
inside air compressed and decompressed		-	-	6 (1)	-	7 (2)	3 (1)	17 (3)
ball deformed		-	3 (1)	-	-	-	-	-
change of shape not mentioned		83 (24)	72 (21)	33 (6)	100 (42)	93 (27)	93 (27)	72 (13)

A sizeable minority, from a third to a half across all Key Stages, gave explanations which made no mention of forces but invoked the round, hollow, rubbery or bouncy nature of the ball itself.

> it is hollow and is made with elastic and rubber

Y6 G M

> the ball has boncey rubber inside it and it makes the ball bonce.

Y3 B H

> It is made of rubber and is round.

Y5 B M

This kind of explanation reveals a focus on some very situation-specific, intrinsic qualities of the object under consideration: in this case, a rubber ball. This kind of response is lacking in generality. It does not refer to the more universal and abstract concepts of force and motion, springiness, elasticity or reaction force

6.3.2.5 Forces on a moving vehicle

Key Stage 2 and 3 children were provided with a line drawing of a moving vehicle and asked to draw and label arrows on it to show the forces acting. For the first part of this probe there was no requirement that the children should consider the *interaction* between the forces they mentioned. It was possible, therefore, for them to think of the forces individually at this stage. The force accounting for the forward movement of the bus was considered separately in the analysis and is reported first. At pre-intervention no child drew a forward-pointing arrow labelled 'engine' or 'wheels', though one third of Key Stage 2 children and a half of those at Key Stage 3 labelled such an arrow pointing in the direction in which the bus was moving 'push'. There was a very small increase in the incidence of such arrows being drawn post-intervention and Key Stage 3 children, in particular, changed to an 'engine' label.

As in some of the probes discussed earlier, it was noticeable that arrows labelled either 'gravity' or 'weight' were lacking in many children's responses. The age-related omissions of any references to gravitational force or weight were in the proportions of two-thirds, a half and one tenth of pupils for Key Stages 1, 2 and 3 respectively. The fact that this discounting of the force due to gravity was apparent in many of the probes used has serious implications for teachers.

Table 6.34   Forces on moving vehicle - from engine

| | **Post-Intervention** | | | |
	KS1 n.a.	LKS2 n=29	UKS2 n=29	KS3 n=18
arrow forwards - engine / wheels		3 (1)	17 (5)	56 (10)
arrow forwards - push		28 (8)	21 (6)	6 (1)
arrow forwards - pull		-	3 (1)	-
no forwards arrow		69 (20)	59 (17)	39 (7)

The problems associated with children's understanding of air resistance were discussed in detail in Section 6.2.3. Their difficulties were clearly revealed in their responses to this moving vehicle probe. There is possibly less confusion on display in this task than there was in the dropped can probe but clearly the use of the words 'wind' or 'wind resistance', with their connotations of moving air, need to be subjected to explicit discussion and reflection in classrooms, in the interests of better understanding of air resistance.

It was not clear in all cases whether children used arrows to indicate the point of action of the force or its direction. Prior to the introduction of the arrow convention to represent forces, many children used arrows as devices to indicate *where* forces were thought to be acting. Such responses cannot be assumed to include ideas about direction, magnitude or movement. Elsewhere in the curriculum, arrows are used in precisely this more limited manner – simply to indicate where objects are, as a labelling device. The probe, as it was presented, might have cued children more precisely to the fact that it was the formal scientific ideas about forces that was the object of enquiry. Equally, such a request might have bewildered those children who did not have such an understanding and it was left to those who did to demonstrate as much. Children's 'friction' arrows highlight this difficulty, for most arrows seem most validly interpreted as indications of *where* friction was thought to be acting, in a rather more general sense than the arrow convention, used more precisely, would enable them to communicate.

The school bus probe was not used, pre-intervention, so data concerning children's awareness of friction can only be compared with other tasks. In these other pre-intervention tasks, friction was not commonly mentioned by children in any of the Key Stages from which the sample was drawn. The friction arrows drawn post-intervention in relation to the moving bus probe occurred with greater frequency than had been the case in other pre-intervention tasks: about one quarter of pupils at Lower KS2, about one half at Upper KS2 and four fifths at KS3. However, the warning that the majority of arrows drawn might have been intended to indicate the *location* of friction rather than the direction in which the frictional force was acting, must be carefully heeded. Arrows were drawn at various angles to the horizontal. Not a single pupil at any Key Stage drew an arrow to represent unequivocally the frictional force acting at the wheel in the same direction as the forward movement of the bus. This is not altogether surprising, since it is a counter-intuitive idea as well as being difficult to reconcile with the notion that might have been promulgated, that friction *opposes* movement. It seems to be safe to conclude that teachers had not addressed this specific issue with pupils, and indeed, there was no imperative that they should have done. The direction of frictional force at the wheels requires a careful and detailed consideration of what parts are moving against what surfaces. Once it is appreciated that the wheels are rotating against the surface of the road, pushing the bus forwards against the purchase on the road surface, it becomes easier to appreciate that the frictional force between tyres and road acts in the direction in which the bus is moving. The example of tyres failing to grip, resulting in spinning on an icy road surface, helps to amplify this understanding.

Figure 6.16

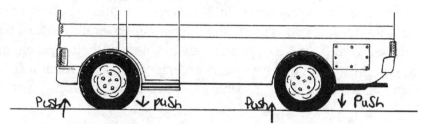

Y6 B L

Pupils were asked explicitly to comment on the changing size of the forces while the bus was increasing its speed.

About one quarter of the KS3 sample suggested that, under the condition of the bus increasing its speed, *all* the forces would be increasing.

They will get bigger and bigger.

Y7 B H

The friction, engine power, and air resistance increase

Y7 B H

About one quarter offered the view that while the bus was increasing its speed, the force attributed to the engine would *increase* while all opposing forces would remain *unchanged*.

the force at the front increases and the force at the back stays the same

Y9 G L

124

Another quarter of the KS3 sample mentioned changes only in the forces opposing movement.

Fiction becomes less and air resistance becomes less

Friction becomes less and air resistance becomes less because you are getting at a faster speed.

Y9 B L

As discussed in earlier sections (see 6.3.1.1 and 6.3.1.2) the evidence suggests that the idea of reaction force is not well developed. In labelling the forces acting on the moving bus, the indications were that pupils had not thought systematically in terms of forces acting in pairs. The moving bus probe confirmed that most children were not thinking about reaction forces, the incidence of reaction force arrows being only ten per cent overall. Most of these references to reaction forces were from KS2 children, suggesting that their teachers had spent some time attending to this aspect of their understanding during intervention. There is thus reason for optimism that KS3 pupils are capable of even greater gains, given appropriate guidance.

Table 6.35   Forces on accelerating vehicle

	Post-Intervention			
	KS1 n.a.	LKS2 n.a.	UKS2 n.a.	KS3 n=18
Balance of forces engine force greater				44 (8)
no size comparison				56 (10)
Changes in size of Forces all forces increasing				28 (5)
engine force increasing, opposing same				22 (4)
engine force increasing, opposing decreasing				6 (1)
all forces decreasing				6 (1)
only opposing forces mentioned				28 (5)
changes in size not mentioned				11 (2)

At the post-intervention stage, KS3 pupils were additionally asked, 'What can you say about the forces on the bus while it is increasing its speed?'. Just under half the KS3 sample indicated that the force from the engine would need to be greater than the forces opposing the forward motion of the bus. The remainder of the sample made no comparison of the sizes of forces. (Section, 6.3.3.1, reviews ideas about the forces acting on the bus while it is moving at a constant speed.)

One lone voice suggested that all the forces would be decreasing. No pupils suggested that all the forces would stay the same, though a minority offered no suggestions about any forces changing in size.

The relationship between forces in the direction of movement and those opposing movement when a vehicle is accelerating is conceptually complex and linguistically demanding to articulate. The force in the direction of movement increases, but so too do the forces opposing that movement, but the increase in the forward force is greater than the increase in the opposing forces. In this sense, the quarter of the sample which asserted that *all* the forces were increasing was correct. The single pupil who suggested that the force of the engine was increasing while the opposing force was decreasing was not literally correct, though it would be true to say that the opposing force was decreasing *relative* to the forward-acting force. The essential idea of the necessity for forces to be *unbalanced* in order for acceleration to be achieved by a moving vehicle was appreciated by less than half of this KS3 sample. It seems likely that pupils would benefit from class debate about the forces acting in various situations such as that presented in the moving bus probe. Such activities would encourage the articulation of ideas and more analytical thinking stimulated by a critical reflection on their own and others' expressed ideas.

6.3.3   Balanced forces on moving objects

It is important to point out at this point that the National Curriculum draws a firm distinction between balanced forces acting on *stationary* objects and balanced forces acting on *moving* objects. While the former are part of the Programme of Study for KS2, the latter are not expected to be addressed until KS3.

6.3.3.1   Forces on a vehicle moving at a constant speed

The probe centred on the subject of the moving bus discussed in section 6.3.2.5 was extended at the post-intervention stage, with KS3 pupils only, to include the question, 'What can you say about the forces on the bus when it is moving at a steady 30 miles per hour?'. (Incidentally, metric units were not used in posing this question in view of the fact that road signs and everyday usage in the UK refers to m.p.h. rather than k.p.h.)

Starting with the essential idea, that of the necessity of appreciating that the forces acting on a body moving at a steady speed must be *balanced,* it is evident that this understanding was not at all well established in the interview sample. Only three children responded in a manner which indicated awareness that, under the condition of constant speed, the forces acting must be balanced (see Table 6.36).

The force that makes the bus move is equal the air resistance and friction

Y9 B M

126

Over half of the sample did not explicitly articulate any relationship which could be interpreted as a view on the balance between the forces causing the bus to move in a forward direction and forces opposing such movement. As such, these pupils did not show awareness of the essential feature of the physicists view of a body moving at constant speed.

Table 6.36   Forces on vehicle moving at steady speed.

	Post-Intervention			
	KS1 n.a.	LKS2 n.a.	UKS2 n.a.	KS3 n=18
Balance of Forces				
forces 'balanced'				6 (1)
engine force same as opposing				11 (2)
forces unbalanced				28 (5)
no size comparison				56 (10)
Changing Size of Forces				
all forces stay same				28 (5)
engine force decreases, opposing stay same				6 (1)
all forces decreasing				17 (3)
only opposing forces mentioned				17 (3)
changes in size not mentioned				33 (6)

Just over one quarter of this KS3 sample suggested that all the forces would stay the same. This response, albeit tacitly, might be inferred to incorporate a notion that the forces which are acting are balanced.

Ch   *The force that makes the bus move is equal to air resistance and friction.*

Y9 B M

Ch   *The engine force will be the same as friction and air resistance.*

Y9 B M

Slightly under one fifth of the sample suggested that all forces would be decreasing.

Ch   The force slows down friction would slow down because the bus is at a steady 30 mph.

It is possible that everyday experience of the sound of the high rate of engine revolutions required to accelerate a vehicle to the nominal speed are being contrasted, by these pupils, with the relatively lower rpm at constant velocity. This is apparent to drivers and passengers alike as the result of the sound of the engine though, or course, the vehicle will be operating in a higher gear. Thus they may conclude that the force to move a vehicle at steady speed is decreasing, or the vehicle is 'coasting'. This may also explain the single pupil who suggested that the engine force decreases while opposing forces stay the same.

The remaining 40 per cent of this KS3 sample either did not mention forces at all, or made reference only to opposing forces.

Under frictionless conditions, Newtonian views of forces become more lucid, their relevance and utility become more apparent. On the surface of the Earth, the Newtonian ideal is veiled; a force has to be provided in order to maintain a movement of any kind. This is because the frictional forces which oppose movement are pervasive, whether it be the result of an object moving across the ground, through the air or across the surface (or below the surface) of water. Pupils lack first-hand experience of frictionless conditions. In situations involving interacting forces such as that of the bus moving at constant velocity, the necessity of balanced forces for a steady speed to be maintained is likely to be counter-intuitive, even at KS3.

6.3.3.2   Stopping a spacecraft in space

It was suggested in the previous section that some of the difficulties children have with envisaging the forces acting on a moving body on Earth are related to their lack of direct experience of frictionless environments. Though Newton's laws of motion predict that an object will continue to move in a straight line in the direction in which a force is applied until something else happens to change that state, this is not the experience on Earth. Moving objects slow down very rapidly as the result of opposing frictional forces. The power of Newton's laws of motion is that they have applicability far beyond the parochial conditions pertaining on the Earth's surface. Indeed, it is on a planetary scale and beyond that the laws are particularly useful, and also, where they can be verified. The modern era of space exploration is not just *helpful* in supporting pupils' understanding of the laws of motion; such a perspective is *essential*. Historically, hypothesis generation and verification of the laws was conducted by means of astronomical observation and it is no coincidence that the period of rapid theoretical advances coincides with the development of the technology of telescopy. In the modern era, images of space exploration, Moon walking and weightlessness are familiar to most children through the secondary source of video material. Children can be invited to engage in thought experiments in which they consider the consequences of various actions in a hypothetical gravity-free, frictionless environment. Children's awareness of and fascination with space travel permitted the use of such a context to probe their understanding of movement and balanced forces. The question posed to KS2 and Ks3 children was, 'If astronauts want to bring their spaceship to a stop in space, what do they do?'.

As expected, children took this question in their stride. No child pointed out that it would be most unlikely that an astronaut should want to bring a spaceship to a halt in space, nor attempted to question what 'stop' would mean in these circumstances.

Table 6.37   Stopping a spacecraft in space.

| | Pre-Intervention | | | | Post-Intervention | | | |
	KS1 n.a.	LKS2 n=29	UKS2 n=29	KS3 n=18	KS1 n.a.	LKS2 n=29	UKS2 n=29	KS3 n=18
reverse thrust		-	7 (2)	28 (5)		10 (3)	10 (3)	28 (5)
create stopping force		3 (1)	3 (1)	11 (2)		3 (1)	-	6 (1)
stop engine		31 (9)	24 (7)	33 (6)		17 (5)	38 (11)	39 (7)
push button		41 (12)	20 (6)	17 (3)		41 (12)	14 (4)	6 (1)
land		10 (3)	-	-		10 (3)	3 (1)	-
stopping impossible		3 (1)	7 (2)	6 (1)		3 (1)	3 (1)	6 (1)
stopping mechanism not known		10 (3)	38 (11)	6 (1)		14 (4)	31 (9)	17 (3)

In order to consider the problem in terms of forces, it was necessary to recognise that most frequently, spacecraft travel through space with rockets or engines switched off, no force being necessary to maintain movement at a constant speed in the frictionless environment of space. Stopping therefore requires a force to be applied in the direction opposite to the movement of the craft.

As might be expected, most children generalised their experience of moving and stopping a vehicle on Earth. The responses of the youngest children in the sample tended not to be able to specify a causal chain of events which would cause the spacecraft to stop: they suggested pushing a button or some similar act of faith.

Ch   They turn the gears off with a switch.

Y3 B H

Ch   Press a button.

Y3 B M

One third of the total sample suggested that the spacecraft could be brought to a halt by 'stopping the engine'.

> turn the engine
> off and then they
> chain it to the plant

Y9 B H

> they turn off the engine

Y9 B H

Four children in the course of the pre-intervention interviews and two during post-intervention demonstrated their awareness that stopping a movement requires a force to be applied, though without being able to suggest a means of applying such a force.

During pre-intervention interviews, a sprinkling of KS2 pupils and a quarter of the KS3 sample gave a 'reverse thrust' form of response.

> To bring them to a stop they put on their rockets
> which fire forward to slow them down.

To bring them to a stop they put on their rockets which fire forward to slow them down.

Y9 B H

The number of KS2 children offering a response of this kind increased slight post-intervention, while the KS3 level of accurate response remained unchanged.

The fact that pupils readily accepted this form of problem and gave serious thought to their responses is encouraging. There are many hypothetical situations which could be discussed by a class which are likely to stretch and enhance their thinking about forces and motion. It is increasingly the case that massive objects need to be manoeuvred in space; occasionally there are accidents which make the news and challenge pupils' assumptions about 'weightlessness' and the relationship between gravity, weight and mass. Such examples are both stimulating and instructive and confirm that science must be an imaginative as well as an empirical discipline.

Summary

The responses of pupils representing the age-range 5-14 years to a wide range of concept probes have been summarised. Shifts in their thinking have been identified and some tentative hypotheses as to why (or why not) ideas might have changed have been very briefly rehearsed. The next chapter reviews the qualitative and quantitative evidence which has been assembled, relates this evidence to the published literature and offers some suggestions for consequent action to support teaching and learning in this area.

7. SUMMARY AND CONCLUSIONS DRAWN

7.0.   Introduction

The programme of research reported here was dynamic and exploratory; it built on previous published data and did not expect to provide the last word on the subject of teaching and learning about forces. It is claimed, nonetheless, that some new insights were gained – about the sequence in the emergence of ideas, about specific teaching strategies which seem likely to enhance the possibility of pupils making progress with their knowledge, about approaches to the notation specific to understanding and communicating ideas about forces and about possible sequencing of the curriculum. These ideas are preceded by some more general points about teaching forces from a position of being informed by a constructivist rationale. It is planned to produce some support materials for teachers following a more exhaustive review of our own and other researchers' evidence and recommendations.

The structure of the chapter is as follows:
    Section 7.1   Some general assumptions
    Section 7.2   The notation of forces
    Section 7.3   Evidence of progression in ideas
    Section 7.4   Some initial thoughts on sequencing

The programme was a demanding one for all participants – researchers, teachers and pupils. In the circumstances, there could only be indications of possibilities for enhancing pupils' understanding arising from particular insights in individual classrooms. In the course of the research itself, there was discussion and reflection, but the possibility of a wide dissemination and implementation of emerging best practices was severely constrained, both by the demanding research schedule and by teachers' wider curricular responsibilities. This report begins the process of reflecting on the insights gained; there was not the opportunity for teachers to re-visit topics, to explore some of the emerging strategies, to attempt to improve on their practices in the light of what had been learned. Such modifications of approach will need to await the next time in which they approach the teaching of forces – for most, the next academic year with the next year's cohort of pupils. At that time, effective elicitation strategies will be familiar and available rather than novel; pupils' ideas might make more sense in terms of the developmental sequences discussed below; possibilities for intervention strategies as means of guiding and supporting pupils' developing ideas will have been rehearsed and prepared. Teachers will be in a stronger position to optimise pupils' progression as the result of the more precise targeting of intervention to pupils' expressed ideas.

The results reported in the previous chapter were described as 'conservative' in the sense that the measured changes in pupils' understanding suggested particular areas in which there seemed to be scope for action. This chapter reviews what we regard as the more significant outcomes of the research which would be expected to carry forward into implications for practice.

7.1   Some general assumptions

*   *Starting early*
    We fully endorse the view of the Waikato group that we must start early:
    '...if children are to understand important ideas in physics it is essential that many
    change their ideas about force. We believe this can be done and suggest the earlier it
    can be done the better; before a child's framework of ideas becomes inflexible. The
    activities booklets...suggests how this might be done with children as young as 11
    years old. However, it cannot be done in isolation from ideas about friction and
    gravity.' (Osborne, Schollum and Hill, 1981, p.21)

    Our only disagreement is that we would not wait until children are eleven years of
    age; there is much that can be achieved in the early years of schooling.

*   *Supporting metacognitive strategies*
    We assume throughout the more specific remarks in the sections which follow that
    children will be encouraged not just to think, but to think about their thinking. This
    means being aware that they have ideas, that their peers have ideas and that scien-
    tists have ideas. It means engaging in the intellectual struggle to articulate unam-
    biguously and consistently their own representations as well as considering serious-
    ly the ideas of others.

*   *Teachers taking children's ideas seriously*
    We assume that teachers will be interested in children's ideas and will take these
    ideas seriously. Recognising pupils' starting points is essential to supporting learn-
    ing with understanding. This 'taking seriously' means accepting them as provision-
    al, accepting the limits on children's understanding, while at the same time helping
    them to develop their ideas as far as they are able. Often, this will imply less,
    (often far less) than conventional scientific understanding; it means accepting the
    principle of 'intermediate understanding' as an educationally valid construct rather
    than a threat to standards of scientific accuracy.

*   *Accepting the refexivity of constructivism*
    Constructivism as a theory applies just as much to university researchers and teach-
    ers as to pupils. Teachers must scrutinise their own understanding of the concepts
    which they are addressing with their pupils.

*   *Seeking evidence for beliefs*
    Science frequently uses empirical enquiry to seek evidence and test hypotheses. It
    is also an imaginative activity, but one in which beliefs are required to be supported
    by evidence. We assume that pupils will be encouraged to test their beliefs against
    primary and secondary sources of evidence derived from and motivated by, whenev-
    er possible, their own active enquiries.

7.2   The notation of forces

Members of a culture share meanings. They extend their spoken communication by using external symbol systems. Some of the notation systems that are relevant to the communication of scientific ideas about forces are arrow notations, drawings, language and quantification. Each of these external symbols systems will be reviewed in the light of the data emerging from the reported research.

7.2.1   The use of the arrow notation to represent forces

Arrows are pervasive in modern society, though in their abstract rather than their physical manifestations. They may serve to remind us of the span of hominid cognition from making, testing and using flint artefacts through to defining the properties and uses of arrows as abstract symbols to represent forces. Arrows as symbols are what concern us here.

Figure 7.1   Some demands of the arrow notation to represent forces.

Drawing demand (Single arrow)	Conceptual demand (Single arrow)
Direction of arrow	Force has direction
Straightness of arrow	Forces act in straight lines
Length of arrow	Magnitude of force
Location of arrow's tail	Objects as point masses
(Two arrows)	**(Two arrows)**
Arrows drawn 'head-to-head' - equal lengths -	Opposing forces - Balanced forces (zero net force, body being stationary or moving at constant velocity)
- unequal lengths -	- unbalanced forces
Arrows drawn in same direction	Forces are additive in same direction
Arrows drawn in different directions	Both forces influence resultant movement
(Multiple arrows)	**(Multiple arrows)**
Arrows drawn in various directions to represent all the forces acting on a body	Directions of individual arrows combined to determine total force; equal and opposite forces cancel one another out. With unequal opposite forces, the net force is the smaller subtracted from the larger.

The research reported above described how pupils showed very little evidence of having been exposed to teaching and learning about the convention of using arrows to represent forces. This understanding was probed with both KS2 and KS3 pupils, pre-intervention, but in view of the results obtained, was not pursued with the younger group in the post-intervention interviews. The KS3 post-intervention results showed a dramatic increase in pupils' appreciation of the arrow convention to represent *direction,* with gains in understanding of the representation of *magnitude* substantial, but not quite of the same order. The interpretation of net direction of movement was not pursued. The evidence suggests that the use of arrow drawing in association with work on forces is under-exploited. Extrapolating downwards from the KS3 results, it seems likely that the greater use of the convention could result in positive gains amongst KS2 pupils also, though this remains to be confirmed. It would seem to be profitable to consider more precisely what the notation *demands* of the learner, and what it offers in the sense of *supporting* understanding. Such a review requires, first of all, some thought about external symbols and notation systems in general. To begin the analysis, the aspects of the drawn symbol and its meaning presented in Table 7.1 seem to be relevant to the age group and the KS1-3 curricular demands, though it is not suggested that all need to be understood at once.

Lee and Karmiloff-Smith (1996) report a lack of consensus in the literature over the technical vocabulary used to discuss notation systems but suggest three major principles: they are independent of i) their creator, ii) location and iii) time. Notation systems operate across generations and facilitate the communication and accumulation of knowledge. It is clear that children are capable of understanding and manipulating a number of notation systems from an early age: drawing, written language, maps, scales, number and musical notation. From the age of two, some understanding of the 'stand for' relationship between notations and what they represent is in evidence, and at three, they can use notations to solve problems in the real world. It is accepted that the pace of development may vary from the use of one system to another, (Lee and Karmiloff-Smith, *op cit*.). In the case of arrow notation, the precise relationship between internal representations and external notation remains to be described in detail. How much of the burden of accurate notation resides in the symbol system itself and how much in conceptual understanding? The answer is that we do not know, but it is possible to offer, even at this stage, a logical analysis of the demands, informed by limited empirical data from our study.

The demands indicated in Table 7.1 are only a beginning of the analysis. For example, even commonplace everyday instances of motion are likely to involve multiple interacting forces. It is also critically important to know, if we are to predict movement outcomes, how such forces are acting in relation to the centre of mass of the objects under consideration. (A javelin will travel in a 'straight' line only if the force is applied though its centre of mass; anywhere else and it will spin, albeit still around its centre of mass.) While acknowledging that we are on the threshold of great complexity, we should not become faint-hearted; the objective is one of helping pupils through a constructive series of intermediate understandings. The way forward is to determine what is accessible to pupils at what age and stage of their thinking and plan teaching accordingly.

If they do nothing else, arrows signal direction, so this would seem to be the appropriate starting point. As indicated in Chapter Six, there is a tendency to use arrows to label *location,* a quite legitimate function in other areas of the curriculum, but not when dealing with forces. Location can be labelled accurately by arrows coming from and pointing towards no matter where, (though there is a more general convention in labelling science diagrams to use horizontal or vertical arrows). Conventional understanding is that which is *agreed*. There seems to be minimal conceptual demand in *agreeing* to use arrows to label the *direction* of forces. Such a resolve is likely to encourage children to think more carefully about forces having direction, and the direction of the particular forces they are representing.

More problematic is likely to be the agreement that *forces act in straight lines*. Children did not see the necessity of drawing straight lines, though a fundamental concept in Newtonian descriptions of force and motion is that forces act in *straight* lines. This is an example of a critical interface between conceptual understanding and conventional representation: children will more likely draw straight arrows if they have it in mind that forces act in straight lines. Is it perhaps legitimate to argue the converse: children will more likely think of forces acting in straight lines if they have been encouraged to draw straight line arrows to represent forces? (We must know our enemy: everyday experience shows us that thrown objects follow parabolic paths through the air; even worse, footballs are intentionally 'bent' around defences and cricket balls 'swing'. Such trajectories have to be understood, in time, as the result of complex interactions of forces.)

Turning to the length of an arrow as a representation of the magnitude of a force, agreement to use the convention seems to be all that is required. (Agreement has both a social and *affective* dimension; in this context, intrinsic interest might suffice.) Our data suggest that young children readily arranged pushes and pulls ordinally, so relative magnitude is not a difficult idea. Of course, more precise quantification and the use of measurement scales will come later.

The idea that the position of the arrow's tail is important can probably be introduced in a macroscopic manner at Key Stage 2, since centre of mass is likely to be a difficult idea in this age group, especially if it involves irregular objects and notions of density.

To represent arrows 'head-to-head' in order to represent forces acting in opposition to one another, or as reaction forces, does not seem to imply any great conceptual burden. The arrow notation might actually help children to make better sense of reaction forces, offering more accessible support than words alone to the formation of the concept that rigid bodies can 'push back'.

7.2.2   Drawing

It is a familiar strategy to most teachers to approach a difficult topic from several angles, different perspectives, using analogies, models and whatever comes to mind to find the representation which 'works', the 'key to unlock the door'. In the theoretical rationale underpinning our approach, we adopt a similar but more formalised view, that of Karmiloff-Smith's Representational Redescription. The research presented in this and previous reports in the series has used children's drawings to illustrate children's beliefs very extensively. These drawings are useful in communicating something of the quality of classroom activities in which children engaged, but they are far more than that, and far more than cosmetic decorations. The extensive use of drawings reflects our view that this form of notation is one which is easily accessible to children as a modality through which their internal representations of how the world works may be externalised. This report, particularly Chapter Five, also reproduces extensively examples of children's drawings. These illustrations usually have comments attached to them, sometimes in the child's hand, at other times annotated by their teacher or the verbatim comments drawn from the individual interview. Many teachers have adopted the technique of annotated drawings as an elicitation strategy. As well as being useful diagnostically, the drawings can be retained for reference as a record of children's thinking at a given time.

We have long recognised that engaging children in elicitation activities cannot be a cognitively neutral activity, any more than it can be affectively neutral. Being asked to articulate one's ideas clearly, consistently and unambiguously, being questioned about details of meaning, however supportively and congenially this is conducted, must be expected to have an impact on thinking. Ideas which might have existed only in the most inchoate intuitive form are required to be explicitly articulated. This is not a problem, other than being an issue which must be honestly addressed in reporting research which might claim to be collecting 'baseline' data. In the context of drawing, the principle of explication can be viewed positively and deliberately as part of the process through which children construct their meanings. There was one striking example which can be interpreted in terms of the impact of the use of drawing on thinking, that of children's drawings of the ball hitting the playground and bouncing away.

In the pre-intervention activity, children were asked to draw the ball in sequence, in the three frames provided. (Previous experience confirmed that children tend to be familiar with comic-strip conventions and are perfectly happy to represent sequential points in time in this manner.) When compared with their post-intervention reponses, it was apparent that there were far fewer references to the deformation of the ball on hitting the ground when the response was elicited independent of the drawn representation. It is inferred that the drawing focused children's attention on the shape of the ball and helped them to frame a more accurate response. Since reaction force seems to present particular problems, this example of the support which drawing can offer is potentially valuable. Of course, drawings could be further annotated with words and arrows.

7.2.3   The language of forces

Language is another example of a symbol system used by a culture to communicate meanings which are independent of particular individuals, time or place. Language is the repository of a culture's knowledge, and the science sub-culture has its own specialist vocabulary which children have to assimilate if they are to share precise understanding of conventional science ideas. A frequent difficulty is encountered when vernacular and scientific vocabulary overlap; Solomon discusses the inherent tensions between the 'life world' and the science domain, (Solomon, 1993). As primary educators, we are perhaps more optimistic that children can be successfully inducted into a more precise use of technical vocabulary. Put more emphatically, accurate consensual language labelling of phenomena of scientific importance must be integral to the acquisition of a scientific mode of thinking. It cannot be regarded as an add-on bonus.

The introduction of the word, 'force' is a good starting point to begin the discussion about vocabulary. Children's examples of forces showed an age-related shift, from concrete, overt actions predominating at KS1 to inferred, abstract instances (for example, of forces acting at a distance) at KS3. Children were asked, 'What name do we use in science for all kinds of pushes and pulls?' in order to ascertain the incidence and extent of the generalised and abstract concept label, 'force'. Only about one fifth of the KS1 pupils generated the word 'force' in response to this question. At KS2, the frequency was about half the sample while at KS3 it rose to around 90 per cent. The research confirmed that the word 'force' is a fairly high level abstraction, one that is not easily accessible to the younger children in the sample. It seems entirely appropriate to guide children towards describing specific events using the terms 'push' and 'pull' at KS1 , as *precursors* to the more generalised term, 'force', rather than as *instances* of the term 'force', bearing in mind that a minority of children may confuse even these simple actions.

Children's understanding and use of some specific terms – 'air resistance', 'gravity' and 'weight', for example, provide further emphatic support for requiring the use of the accurate meanings of words to describe unambiguously agreed phenomena. This is not an argument for teaching vocabulary *independently* of concepts; it is an argument for demanding the correct words to label achieved understandings, so that understanding is maintained and reinforced by correct usage. Some of the vocabulary relevant to forces which would benefit from clear usage is briefly reviewed.

Air resistance. The sensation of a moving body of air on a person – i.e. what is referred to as 'the wind' in everyday expression – is invoked by many children to explain the effect of 'air resistance' as perceived when a vehicle moves at speed through the air. This phenomenon tends not to be understood as a force which opposes the movement of an object which is moving through air. Rather, the 'wind' or 'wind resistance' force seems to be thought of as coming into operation when certain critical thresholds are passed. For example:

- slow moving objects are commonly not regarded as encountering (or generating, as some children would have it) 'wind', 'wind resistance' or more accurately 'air resistance';
- the mass of an object is regarded as critical by many children, so that this force opposing movement is not thought to apply to objects having a large mass;
- the size of an object may, like its mass, be regarded as a threshold property rather than a variable property.

The 'wind' is certainly capable of exerting a force as masses of air shift between high and low pressure areas. Equally, 'air resistance' is a tangible force which opposes the movement of objects through a mass of air. The term 'wind resistance' is one to be discussed if and when it arises, to be subsequently discouraged. It is the result of a conceptual short-circuit between two conceptually discrete areas. There is consequently a danger of a conceptual confusion being cemented by an inaccurate linguistic labelling.

Rather than being thought of as acting within a system of forces, these attributes of moving objects which are deemed relevant to a consideration of air resistance are thought of as *causal* rather than *interactive*. Thus pupils refer to 'wind' or 'wind resistance' – the move-

138

ment of a body of air opposing the direction of movement of an object – rather than an interaction between the surfaces of that object and a body of air.

Gravity. To the scientist, 'gravity' is not a word that carries meaning. In vernacular usage, it probably means something like 'that which holds us to the Earth' – a property of the Earth rather than a force which applies *between* masses, anywhere and everywhere. The everyday definition is actually at odds with the scientific view and probably tends to reinforce an erroneous idea.

Weight. Scientists also use a precise definition of weight, one that is underpinned by assumptions about how gravity acts. Thus, to the scientist, the force on an object due to the gravitational pull of the Earth is what physicists call that object's 'weight'. It was clear in our study that many children were operating a definition of 'weight' which did not take into account a causal relationship between an object and the force of gravity on that object. At the extreme, pupils treated weight and gravity as quite separate. Indeed, some suggested that weighty boots could compensate for the (assumed) absence of gravity on the Moon.

There are wider issues about understanding of forces which are associated with language. For example, the transitive and intransitive use of the verb, 'to move' is discussed in earlier sections. The counter-intuitive sense of 'the wall pushes back' is a difficult enough idea perceptually and conceptually, but one which actually seems to be confounded by the particular language used to describe how reaction forces operate. It might be helpful to substitute another, more acceptable phraseology.The vocabulary has to accommodate (or recognise) the difficulty of sentence construction with respect to the misleading introduction of the notion of *sequence*. 'The wall pushes back' or 'The stool pushes back' is such an unusual use of language that it seems to contradict common sense. 'The wall exerts a force in the opposite direction', might be more acceptable.

The 'narratives' of how force and motion are undertood from different perpectives and belief systems (Appendix III) are offered as a reminder that the 'stories of forces' should not be ignored. Narrative description is a highly accessible modality to children, through which they might be encouraged to relate causal sequences to one another, for articulating and cross-checking of one another's interpretations of events.

7.2.4   *Quantification of forces.*

Quantification is a technique which is fundamental to scientific thinking and enquiry; it is what allows comparisons to be made and results to be accurately recorded and communicated. While in mathematics, the introduction of any physical quantity tends to be carefully graded, there was little precedent of which the research group was aware of a parallel analysis of the introduction of the measurement of *forces*. Following some exploration of pupils' classification of forces, their thinking was later directed towards comparisons of magnitude. This was approached by asking them to name three pushes (and then, three pulls) in order of magnitude. The youngest children were encouraged to draw their responses while the older pupils were invited to make written responses.

At first inspection, the results were surprising in that the performance of KS1 and Lower

KS2 children showed the expected trend of increasing capability with age, while the Upper KS2 and KS3 performance was lower. A scrutiny of responses revealed that the KS3 pupils tended to name formal forces but failed to differentiate these unambiguously in terms of relative magnitude.

It was assumed that a helpful precursor to the introduction of standard units of measurement of force would be for pupils to explore, initially, the quantification of forces using their own non-standard measures. (This is standard good practice preceding the introduction of formal units of measuring length, etc.) In the event, few teachers actually managed to implement the idea of children constructing their own, non-standard force measurers. Successful identification of the commercial force-meter was fairly widespread and showed some increase following the intervention activities. Nonetheless, awareness of its force-measuring function was known to only around half the pupils at KS2 and about three quarters at KS3.

Intervention activities seemed to have been successful in getting pupils to become aware of the units which force-meters measure: 'force' or 'newtons' was offered by about one quarter of Lower KS2 pupils, one third at Upper KS2 and about three quarters at KS3. As regards awareness of the horizontal and vertical possibilities of measuring forces, an appreciation of vertical uses was more widespread though many of these referred to measuring *'weight';* horizontal uses were mentioned much less frequently, the highest rate in the groups questioned being about one third of Upper KS2 pupils.

In summary, the force-meter's use is neither well established nor well understood at KS2 and KS3 and the widespread implementation of this aspect of the curriculum remains to be achieved. It is not suggested that measurement of forces *per se* is of paramount educational value, rather that quantified values offers another way of thinking about, manipulating and 'redescribing' aspects of forces. Such opportunities need to be exploited as they arise. For example, the simple exercise of ordering everyday events, initially as an ordinal series, perhaps moving to estimating the absolute values of the forces involved in newtons, and finally to measured comparisons, would be the sort of sequence likely to prove useful. Current practice seems to miss out the stage of offering pupils opportunities to quantify intuitive experiences.

7.3   Evidence of progression in ideas

7.3.1   Progression in ideas about agency

There was some evidence in the examples children cited when asked to give an example of a push by a non-human agent of the following transition:

1.   The youngest children (KS1) think of themselves and other people as capable of pushing and this capability is generalised to other living things. They tended to think of movement as subjective and active, 'to move' in its intransitive usage. This subjective, active and egocentric view does not necessitate a view of forces as acting between two (or more) objects.

2.   Lower KS2 children revealed more awareness of pushes (and to a lesser extent,

pulls) as being events happening in the natural world. The pushing of the wind was frequently mentioned, and other geo-physical events to a lesser extent. Appreciation of the range of non-living instances of phenomena which are capable of exerting a force might be important in helping children to shift from a subjective to more objective conceptualisation of agents, though some carry over of animistic attributions is well-established in the literature.

3.   The responses of pupils at Upper KS2 contained more examples of human artefacts in the form of machines and wheeled vehicles than of any other response category. Such examples may be useful in bridging between the intransitive and transitive use and understanding of the verb 'to move'. Vehicles move themselves, but they are also frequently associated with moving other objects.

4.   The main category of response in evidence from KS3 pupils were those labelled by technical terms such as 'air resistance', 'friction', 'gravity' and 'upthrust'. These examples mark a shift towards awareness of technical vocabulary naming forces in more abstract terms. The kinds of examples cited enhance the possibility of forces being recognised as invariably involving pairs of objects.

This sequence is more than a description of a developmental trend. It is an interpretation of the development of pupils' outlook in terms of the factors which impinge and are likely influences on their thinking. The factors which have been selected as salient in this analysis are those which support development in the direction of conventional scientific understanding which holds that forces have to be understood as working in pairs, between objects. The value of mapping such a progression is not just for its intrinsic developmental interest in descriptive and interpretative terms. In an educational context, we seek such developmental predispositions for prescriptive and didactic purposes. If this is the way children's understanding is disposed to grow, we need to ask questions as to how such 'growing conditions' may be optimised in the classroom. In other words, effective intervention should seek to exploit developmental predispositions.

7.3.2   Ideas about the gravitational force of the Earth

While it is established in the literature that pupils may think of gravitational force as either a push or a pull, it was a surprise to find that a significant proportion at all ages (though declining with increasing age) could think of gravity as *both* a push *and* a pull towards the Earth. 'Pull' was the more commonly held view, being expressed by about one third of Lower KS2 pupils, about half at Upper KS2 and four fifths at KS3. It is tempting to infer from these cross-sectional data that the understanding emerges fairly steadily over the seven years between the ages of seven and fourteen and furthermore, that this understanding might be expected to be capable of being accelerated, given focused intervention. About one fifth of KS2 pupils described the gravitational force of the Earth as a 'push'. The proportion of pupils describing the effect of the earth's gravity as '*both a push and a pull*' actually increased, following intervention, to 45 per cent at lower KS2, 17 per cent at Upper KS2 and 11 per cent at KS3.

Very few pupils described gravitational force as an attraction between masses (one Upper KS2 and one KS3) or between the Earth and other objects (one Lower KS2 and two KS3). (See confirmatory evidence from Bar *et al.*, 1997). This understanding was achieved by only a small minority, but might exemplify what many others might be capable of understanding rather than being symptomatic of precocious insight on the part of the pupils involved. Newton's insight is a description rather than an explanation. (The inclusion of variables such as mass and distance and the inverse square law add complexity, but these ideas are not essential to an initial, basic understanding of gravitational force.) Those pupils who offered this most sophisticated level of response are unlikely to have generated such insight through their own individual activity. It is far more probable that the information was socially transmitted. If so, they must have been in a state of 'readiness' for such knowledge. To decide whether it is appropriate for such transmission to be more widely promulgated, we need to know the nature of such readiness. (In Vygotski's terminology, the defining characteristics of the 'zone of proximal development' which makes the learner receptive to the scaffolding of knowledge about attractive force between masses). Logical analysis suggest that an appreciation of the Earth as a separate spherical body capable of attracting objects towards its centre from any point around it is essential antecedent knowledge. Perhaps knowledge that other bodies can have an effect on the Earth – the Moon's effect on the oceans being an example – might also be prerequisite. Hypotheses for intervention such as these, arising from empirical enquiry, need to be fed back into an iterative process of curriculum research; we have to check the circumstances in which teachers can support (or even accelerate) progression. The fact that some pupils have achieved understanding alerts us to the possibilities of others following the same sequence. (We must value pupils' individuality and special talents, but most education is a process of learning the well-beaten pathways.)

Reviewing pupils' understanding of gravitational force, it is possible to suggest four levels of understanding revealing increasingly generalised understanding.
1. Gravity is not associated with any clear direction. Gravity is understood as something which causes objects to fall 'downwards' or towards the ground.
2. Gravity is thought of as a force acting between the Earth and other bodies near the Earth.
3. Gravity is linked to the mass of the Earth and the mass of objects attracted to the Earth.
4. Gravity is conceived as a force between masses which might happen anywhere in the universe.

As we (and various researchers before us) discovered, there are many conditions which pupils see as *variables* impinging directly on gravitational forces which scientists or educators might prefer to describe as *context* effects. To pupils, these conditions are perceived as being causal rather than incidental. Educational research has a complex task to unravel these pupils' perceived effects from their entanglement with the actual effects defined by scientists. The following examples illustrate some of the situational effects which were encountered.
- The Earth's gravitational force is caused by spin or air; gravity might be a pull or push, or both
- Gravity is *not considered to be acting at all* by a significant proportion of pupils when a ball has been thrown and is moving vertically upwards. A minority suggested that gravity would be operating, but to a *reduced extent* during upward movement. When that ball is at the apex of its trajectory, most children suggested that gravity would be operating.
- When a can was thrown from a moving car, forces other than gravity appeared to dominate children's thinking.

- Many children explained the outcome of Neil Armstrong's hammer and feather experiment on the Moon (both objects hitting the ground simultaneously) in terms of an absence of gravity.
- Weight and gravity were treated by many pupils as separate phenomena, leading them to suggest that heavy boots were needed on the Moon to compensate for the lack of gravity. A similar idea was expressed to explain a helium balloon floating in air; paper clips were said to be contributing 'weight' (rather than attributing the downward force to gravity) which was counteracting the tendency of the helium balloon to rise.
- The relationship between mass, weight and gravity was poorly understood in the situation in which pupils were asked to explain the difference between units of measurement of mass and force.

7.3.3   Ideas about frictional forces

Familiarity with the word 'friction' and the idea which it describes were surprisingly extensive. Large shifts were recorded in the direction of an increasing frequency of correct responses, post-intervention, suggesting that this is an area in which gains in understanding may be expected from targeted intervention. A common idea is that friction is acting only when there is *movement* between surfaces, a belief which falls short of the scientific definition in which friction can cause a system to remain static by opposing a tendency to movement.

7.3.4   Balanced forces

A helium filled balloon provided a situation in which pupils might offer confirmation of the perception of the idea of balanced forces. Children added paper clips to the balloon's string until it moved neither upwards not downwards. They were then asked to comment on the forces acting on the balloon. A large proportion of pupils referred to two balanced forces acting with a steady increase in the incidence of correctly identified balanced forces up to about one third of the KS3 sample. (Rather more KS3 pupils referred to balanced forces but did not identify correctly the forces involved; the KS2 sample were more likely to omit any mention of balance or to nominate only one force.

The helium-filled balloon seems to be a particularly useful stimulus to the consideration of balanced forces in a static situation. It also invites analogies with floating and sinking of objects in water.

7.3.5   Reaction forces

A concept probe using a top-pan balance was selected on the basis that this would make the concept of reaction force most perceptible and tangible to pupils. Nonetheless, the conceptualisation of a force pushing back was not obvious or accessible to a majority of KS2 pupils. About 40 per cent of KS2 pupils demonstrated an understanding of reaction force being equal, this proportion rising to about 70 per cent at KS3.

The idea of reaction force was also examined in the context of children's experience of sitting on a chair. About one quarter made reference to 'reaction force' post-intervention, the idea having been scarcely in evidence at all in the same context prior to intervention. A very small number of pupils framed their responses in terms of balanced forces, more nominating incorrect than correct forces. The KS2 pupils performed at least as well as those at KS3. It seems that children may be exposed to several (in this instance, perhaps up to nine) years of teaching about forces without establishing the fundamental idea that forces always act in pairs. Balanced forces in static situations perhaps need to be discussed earlier than is the case in current practices.

Although the situation of a rubber ball dropped onto a playground was anticipated to maximise the possibility of those who had some awareness of reaction force mentioning it – because of the observable compression of the elastic ball – this, in the event, did not occur. Only a very small minority of pupils referred to 'reaction force' either in formal language or via some equivalent circumlocution. Younger pupils tended to centre their attention on the factor of the *hardness* of the playground (one third KS1, one quarter Lower KS2). The appropriate language, by means of which younger children might be enabled to discuss reaction forces in a more meaningful way, is in need of attention.

The pre-intervention elicitation technique was to ask children to draw the ball just before hitting the playground, at point of impact, and in the air bouncing away again. The drawings revealed an appreciation that the ball deformed and then reformed on the part of one fifth of KS2 pupils and two thirds at KS3. In contrast, post-intervention, which invited verbal responses without recourse to drawing elicited a markedly reduced attention to deformation of the ball: less than ten per cent at KS2 and 28 per cent at KS3. The use of drawings would appear to encourage Representational Redescription and the focus on two-dimensional visual representation succeeds in drawing attention to a salient feature in a manner that a verbally articulated response did not.

A large proportion of attempts to explain the bounce of the ball were framed in terms of intrinsic qualities of the ball - its roundness, bounciness, the fact that it was made of rubber – rather than using expressions to describe force and motion. The obvious question for a teacher is how children might be moved from the specific to a consideration of the more general and abstract. One obvious response is that the teacher must encourage children to focus on more general properties. Children might be asked to check other properties of bouncing objects. Do wooden balls, square or solid objects bounce? Maybe posing such questions is sufficient to spur many children to shift to a level of abstraction on the spot, while for others, a longer journey might be expected.

7.3.6   *Momentum*

It is instructive to look back to, 'Toward Changing Children's Ideas' about forces emanating from one of the earliest systematic enquiries in science education which was based on a constructivist rationale: Roger Osborne's 'Learning in Science Project', at the University of Waikato, New Zealand. (See Osborne, Schollum and Hill, 1981; Schollum, Hill and Osborne, 1981.) It is reasonable to ask to what extent teaching approaches towards forces have changed in the almost two decades since the Waikato group's publications. Osborne et al. took the view that children's frequently asserted intuitive view that force is something in a projected object is similar to the physicists' concepts of momentum. (Ogborn and Bliss,

144

1993, demonstrate the extent to which children's 'common-sense' theories of motion, including the idea of *effort of the motion of an object*, can cohere; see Appendix IV.) Physicists do not consider momentum to be a force; it is the product of *mass* times *velocity* of the body, a measurable quantity which can be changed by the application of forces. While, in the physicists' terminology, children would be quite wrong to suggest that a ball thrown through the air falls to the ground because it has used up the force it carried within it, they would be much closer to the conventional view if they used the language of the ball's 'momentum decreasing'.

Pupils' ideas about moving bodies have much in common with the scientific concept of *momentum,* the quantity of motion in a body given by the product of mass and velocity. The points of correspondence are often dismissed by the conventional scientific viewpoint which sometimes refers to them as evidence of a 'naïve impetus theory'. Others suggest a more radical approach, but one which is firmly located in the constructivist rationale of starting teaching from pupils' existing ideas. Since many pupils think of a force as something in a moving object by virtue of its motion, Osborne *et al.*, suggest offering them the correct scientific label for the attribute which they have identified, namely *'momentum'*. Examples are then discussed of objects gaining or losing momentum, while the distinction between force and momentum continues to be identified.

Although validating pupil's notions of momentum does not appear to be a strategy which has gained either widespread approbation or implementation, others who have seen the merit of the approach have taken the idea further. It is possible to conceptualise force as a substance-like quantity having extensive properties, or even as currents of momentum, the mathematical formulation being redescribed as a fluid metaphor. The curricular implementation of such an approach might start with the human understanding of bearing a load, feeling the force required in the opposite direction as an analogy of how the pillar holding the beam bears a load. In static bodies, the tension can be described as the flow of momentum. In dynamic situations, the analogy can be drawn of two buckets linked by a pipe, fluid flowing into one flows into the other. Such ideas are neither fanciful nor lacking in rigour, as the publications of Hermann *et al.* (see for example, Herrmann and Schmid, 1984) testify. The model, it has been suggested, can be adopted with mathematical rigour and consistency to advanced theoretical and applied levels of physics. One of the great attractions of the approach is the link which it encourages between the laws of motion and thermodynamics.

7.4 Some initial thoughts on the sequencing of the teaching of forces

7.4.1   General Understanding of the concept of 'Force.'

It is generally accepted that the learning is more effective if the scientific label for a concept is not given until after some understanding of the concept has been achieved. The evidence from this research suggests that 'force' is no exception to this view point. Nevertheless, a sequence of development towards the correct use of the term force is more likely to be effective if it reflects the cognitive development of the learner. Such a sequence would appear to be:

- whole-body experiences of pushing and pulling (personal to the learner but also recognising the need for an object to be pushed or pulled);
- description of these experiences using appropriate language ('push' and 'pull' extended to include such terms as kick, throw, jerk); such language necessarily implies appreciation of direction;
- extension of 'push' and 'pull' experiences to a recognition of similar actions by non-human animals;
- consideration of the effects of pushing and pulling on the movement and/or shape of the pushed or pulled object;
- using the awareness of these effects as an introduction to pushes and pulls exerted by inanimate objects (cars, magnets, water, wind), with a reiteration of the need for there to be other objects which experience the effects;
- introduction of the scientific use of the term 'force' to cover all forces of push and pull. (One object exerts a force on another).

In terms of any proposed changes to the National Curriculum for England and Wales this sequence would postpone the use of the term 'force' until KS2.

7.4.2   Quantification of forces

Quantification is an essential element of science in that is often contributes significantly to the quality of the evidence being obtained.

The following sequence for the quantification of forces would need to be run in parallel with that suggested above for the development of the general concept of force.

- Description of the size of pushes and pulls in broad terms, (big, small, medium, linked with differing sizes of effects);
- sequencing of given pushes and pulls in order of magnitude;
- meaning of forces in non-standard units (requires the construction by the learner of a suitably accurate force-meter which is capable of measuring both pushes and pulls);
- introduction of the standard unit - the newton (awareness of the magnitude of the newton to be gained from standardisation of the force-meters, estimation exercises and direct measurements).

In terms of the National Curriculum the first two steps in the above sequence would appear to be appropriate for KS1 and the latter two for the second half of KS2.

7.4.3   Conventional representation of forces with arrows

Considerations of the precise meaning of the terms 'push' and 'pull' lead inevitably to the recognition that forces have direction. Children's drawings of situations involving forces often, therefore, include arrows. However, in other areas of the curriculum arrows are used to indicate location. Teachers wanting to use 'force' arrows on drawings in order to help children work through their thinking and hence to assist in formative assessment of understand-

146

ing, need first to ensure that the children recognise the dual function of arrows, namely to represent direction or location. The conventional use of the shaft length to communicate the relative sizes of forces can be introduced at a later stage, subsequent to the introduction of measurement by force-meter.

The use of arrows on drawings to represent both size and direction can be particularly useful at the stage when the children are considering several forces acting simultaneously. (See 7.4.5 Balanced and unbalanced force).

The evidence from some of the schools involved in this research is that, particularly at KS2, children represent 'pushes' with arrows where the head touches the object being pushed and 'pulls' with arrows where the tail touches the object being pulled. It may well, therefore, be advantageous for teachers to use this same modification of the convention in order to communicate ideas to children or to assess their understanding of them.

7.4.4   Specific forces - gravity, friction, air resistance, reaction

Although the implications for the teaching of these specific forces are considered separately below it is assumed that they will, at least to some extent, be taught concurrently.

7.4.4.1   Gravity

Gravity is such a commonly used 'term' that children have it within their vocabulary from a relatively early age. Their appreciation of the meaning of the term develops as more evidence becomes available to them. For teachers helping to provide this evidence and encouraging the discussion of it the following sequence of development is suggested.

- Gravity keeps things on the ground, stops them floating away.
- Gravity is a property of the Earth, so is a pull from beneath.
- The pull of gravity is directed towards the centre of the Earth.
- The size of the force of gravity depends on the mass of the object being pulled by the Earth.
- The size of this force is the weight of the object.
- Gravity on the Moon is less than that on Earth.
- The size of the gravitational force is determined by the mass of the object and that mass of the Earth/Moon/planet.
- There is a gravitational force between any two objects.
- The size of this gravitational force depends on the distance between the objects.

An understanding of the effects of gravity on falling objects does not appear to be directly linked to the developments outlined above. By the end of KS2 most children are willing to accept that the rate of fall is independent of mass but an understanding of why this should be so appears to be beyond most, even at the end of KS3 (See 7.4.3).

7.4.4.2 Friction

The suggested sequence for the concept of friction is that the children are moved from the understanding of 'grip' as a property of an uneven surface to an appreciation of friction as a force acting to prevent relative movement of two surfaces in contact.

- A rough surface impedes the movement of an object across it.
- Different surfaces impede this movement to different extents.
- Introduction of 'friction' as the scientific term for this force which changes movement. (Needs to follow recognition that all changes in movement require the action of a force).
- The direction of the frictional force is opposite to that of the movement.
- Both of the surfaces in contact contribute to the magnitude of the frictional force.
- Friction can occur even in the absence of movement.

7.4.4.3 Air resistance

If air resistance is to be presented as one example of a frictional force, one in which one of the surfaces is fluid, then its introduction would need to follow the sequence outlined in 7.4.4.2 Friction above.

A helpful comparison would be with the frictional force between two surfaces one of which is static.

It is unreasonable to expect children, who do not understand that the rate of fall of objects is independent of their mass, to interpret correctly the effects of air resistance on falling objects.

7.4.4.4 Reaction forces

The conceptual difficulty presented by the concept of reaction force would suggest that it be left until quite late in the overall sequence of the teaching of forces. However, it is again suggested that the teaching begins with whole-body, personal experiences and moves towards the abstract and inanimate.

- Whole-body experiences to consolidate idea of no movement resulting from equal and opposite forces.
- Whole-body actions against flexible objects (springs) to experience 'opposite' nature of reaction force.
- Whole-body action on top-pan balance to introduce equivalence size.
- Inanimate objects on top-pan balance to consolidate equivalence of size.
- Consideration of inanimate object on rigid surface (no movement, therefore balance of forces).
- Consideration of reaction forces in non-static situations.

The arrow representation of forces can be of considerable assistance during discussions of reaction forces. The use of language which does not invoke personal, animate experiences is also to be recommended.

7.4.5   Balanced and unbalanced forces

The consideration of forces in isolation is a useful introduction but eventually it will be necessary to consider real situations in which several forces have to be considered together. The essential pre-requisite is the appreciation that forces have both size and direction.

- Whole-body experiences to recognise that two forces can either oppose or reinforce each other.
- Introduction of the terms balanced (net force zero) and unbalanced for forces in combination.
- Investigations of the effects of forces acting together.
- Continuous application of a force results in continuous change (acceleration).
- Consideration of situations in which there is no change in movement - static and constant speed (balanced forces).

LIST OF APPENDICIES

Forces Research

APPENDIX I

SPACE SCHOOL PERSONNEL

(for research carried out into children's
understanding of Forces in 1996/1997)

School	Head Teacher	Teachers
Bradshaw CP School	Mr John Kenyon	Mrs Jenny Boyle Mrs Kate Dean
Chesterfield High School	Dr Alan Irving	Mrs Joanne Walker
Cole Street Primary School	Mrs Gail Webb *	Mrs Jo Hall
Farnborough Road Infant School	Mrs J Hartsham	Mrs Jayne Haines
Formby High School	P G Baldock	Miss Lilly Eaves
Kew Woods School	Mr DWT Hughes	Mrs Claire Hardy
Park View Primary School	Mr Adams	Mr David Nieman
Scarisbrick CP School	Mrs Sue Harrison *	Mrs Susie Haden Mrs Audrey Stocks
St Andrew's RC Primary School	Mrs E A Jones	Mrs Jean Fitzsimmons
St Lawrence JMI	Mr K Allen	Mr Mark Thomas
St Margarets C of E High School	Dr Dennison	Ms Gillian Shilton
St Oswald's School	Mrs Margaret F Ellams	Mrs Vivian Ward
Windlehurst CP School	Mr Ashcroft	Mrs Gillian Green
Wolveram CP School	Mrs Beryl Clarke *	
Woodend Primary School	Mr Alex Blythin	Mrs Wendy Grime

* Headteachers who contributed to project teaching.

APPENDIX II

EXAMPLES OF CONCEPT PROBES

The following concept probes are those for which data are presented in the main body of the report. They were originally presented in association with practical activities, where possible. They were also bound into Key Stage specific booklets. To avoid repetition, each concept probe is reproduced only once in this appendix. The key on the right indicates by shading the Key Stage to which each probe was exposed.

1.   What do you have to do to make your bicycle start moving?

KS1	KS2	KS3

2.   In the drawing the child is just letting go of the can. The car is being driven forwards quite fast.

KS1	KS2	KS3

a)   On the drawing put a letter C where you think the can **will first** hit the road.

b)   Why do you think the can **first hit** the road at the place where you put the letter C?

152

3.   Think about what you do when you push.
        Draw or write **FOUR** things that you do which are
        pushes.

KS1	KS2	KS3

4.   Think about what you do when you pull.
        Draw or write **FOUR** things that you do which are
        pulls.

KS1	KS2	KS3

5.   Write how you decide whether you are pushing
        something or pulling it.

KS1	KS2	KS3

        A push is when I ....

        A pull is when I ....

6. (a) Can you think of a push that is **NOT** done by a person?

KS1	KS2	KS3

    (b) Can you think of a pull that is **NOT** done by a person?

KS1	KS2	KS3

7. Can you think of a **very small push**?
    Draw or write your idea in the box.

    Can you think of a **very big push**?
    Draw or write your idea in the box.

    Can you think of a **medium-sized push**?
    Draw or write your idea in the box in the middle.

KS1	KS2	KS3

a very small push	a medium-sized push	a very big push

8. Can you think of a **very small pull**?
    Draw or write your idea in the box.

    Can you think of a **very big pull**?
    Draw or write your idea in the box.

    Can you think of a **medium-sized pull**?
    Draw or write your idea in the box in the middle.

KS1	KS2	KS3

a very small pull	a medium-sized pull	a very big pull

154

9.   What words do we use in science for all kinds of pushes and pulls?

....

KS1	KS2	KS3

10. (i) What it this measurer called? ....

(ii) What does it measure? ....

(iii) Give **TWO** different ways that it can be used to measure.

a) ....

b) ....

(iv) Rulers measure in centimetres.

This measurer measures in ....

KS1	KS2	KS3

11. (i) Is the effect of gravity on objects a PUSH, a PULL or BOTH.

Tick **ONE** box to say what gravity is.

a push   ☐

a pull   ☐

push and pull   ☐

(iii) Explain how gravity works.

....

....

....

KS1	KS2	KS3

12. A ball is thrown up in the air.

KS1	KS2	KS3

Is gravity acting on the ball:

(i) when it is moving upwards? ....

(ii) just when it reaches its highest point? ....

13. Why do astronauts wear big boots when they walk around on the Moon?

KS1	KS2	KS3

....

14. Mary has a 500 gram packet of butter.
When she hangs it on a newton-meter it reads 5.

KS1	KS2	KS3

Why does it read 5, and not 500?

....

15. Jim rides his bicycle across the school playground.
When he rides across the school field he has to pedal harder.

KS1	KS2	KS3

Why is it harder to ride across grass?

....

....

156

16. Under each drawing write if you think that the force of friction is acting on the book.

Plank level
book still

....

Plank raised a little
book still

....

Plank raised higher
book sliding

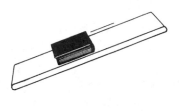

....

KS1	KS2	KS3

17. In the video you saw the astronaut on the Moon drop a hammer and a feather. They both hit the surface of the Moon at the same time.

Why does this happen on the Moon but not on Earth?

....

KS1	KS2	KS3

18.   John pushes down on the scale pan with his hand.

The reading on the scale pan is 10 newtons.

What will be the force on his hand?

Tick **ONE** box.

Reading on force-meter	
less than 10 newtons	
10 newtons	
more than 10 newtons	

19.   What forces are acting on you when you sit on a stool?

        ....

        ....

20.   If you drop a tennis ball onto the playground it will bounce.

        What makes it bounce?

        ....

        ....

158

21. If astronauts want to bring their spaceship to a stop way out in space, what do they do?

KS1	KS2	KS3

....

....

22. The children have put some paperclips onto the string of their helium balloon.
Their balloon stays still. It does not move up or down.

KS1	KS2	KS3

What can you say about the forces acting on their balloon when it is like this?

....

....

23. Look carefully at the drawing

What exactly do the arrows on the drawing tell you about the forces acting?

KS1	KS2	KS3

....

....

....

....

....

....

23. The school minibus is moving along the road.

(i) Draw arrows on the picture to show the forces acting on the bus.

(ii) Put labels on the arrows to show what the forces are.

(iii) What can you say about the forces on the bus while it is increasing its speed?

KS1	KS2	KS3

....

....

(iv) What can you say about the forces on the bus when it is moving at a steady 30 miles per hour?

KS1	KS2	KS3

....

....

APPENDIX III

THEORIES OF MOTION

Aristotle's cosmology

'Aristotle's treatment of how bodies move was one of the characteristic determining features of his world picture or cosmology. Heavy bodies move downwards in a straight line towards the centre of the universe, which is the centre of the Earth; light ones, that is, bodies which have positive lightness, move away from the centre, again in a straight line. This is the key to the doctrine of the elements. Earth goes straight down; fire goes straight up; air goes up because it is light, but not as light as fire; and water goes down, because it is heavy, but not as heavy as earth. So we have earth, surrounded by water, then air and finally fire: these four spheres make up the elementary world of below the Moon. The only bodies that are neither heavy or light are by the same token non-elementary: they are heavenly bodies with their own appropriate motion, which is circular round the centre of the universe. Circular motion is appropriate to heavenly bodies since no change has ever been observed in the heavens. All we see is endless repetitions of the same patterns of movement, but there is no trace of the generation and decay which is the mark of our elementary world. The heavens must be made of a fifth element (quintessence), an imperishable, incorruptible substance. Since the quintessential heavens are completely different from the Earth and its surrounding elements, there could be no thought of treating all motions in the universe as subject to the same laws. As far as local motion on or near the Earth was concerned, Aristotle was content with principles that were more or less satisfying to commonsense, at least until subjected to a serious examination.'

'There was no need to explain why a body was at rest in its natural place: that was where it was supposed to be, so it could not be expected to move from there unless forced to. Physics was the study of nature: central to it was the study of natural motions, the study of how bodies return to their proper places. It was motion, not rest, that needed an explanation. There were, or course, also motions that were not natural: these were forced or violent motions. Things like chairs or carts or spears did not move of their own accord: to move them took effort, which was to be expected because they were being moved from their natural rest; violent motions merited incidental attention. But even with natural motions like the free fall of heavy objects, the resistance of the medium had to be considered, since a stone, for instance, obviously falls more quickly through air than through water. In fact, Aristotle took heavy bodies to fall with speeds proportional to their weights in a given medium. He also took the speed of fall to be inversely proportional to the resistance of the medium, though what Aristotle had in mind is more faithfully captured by Galileo's terminology: the more subtle the medium, the faster the body falls; the crasser, the slower.'

From: Michael Sharratt's (1994) 'Galileo. Decisive innovator. Blackwell, Oxford UK and Cambridge USA. pp.30-31.

Newton's laws of motion

- Every object stays at rest or in a state of uniform motion in a straight line unless a force acts upon it.
- If a force acts on an object, that object accelerates at a rate given by dividing the force by the mass of the object.
- If one body exerts a force on another body, then the second body exerts an equal and opposite force on the first.

Sketch for a common-sense theory of motion

From: Bliss, J and Ogborn, J, (1993) A common-sense theory of motion. Chapter 7, pp. 120-133. In Black P. J. and Lucas, A. Children's informal ideas in science. Routledge. London and New York.

Two basic and related terms of the theory are 'support' and 'falling'. If an object is supported, it does not fall; if it is not supported, it falls, until it is once more supported. Falling has an initial cause, namely a loss of support, but is a natural motion in that one need not look for a cause (a force or agency) for it to continue, only for a continued lack of support.

Everything needs support, except only the ground, which gives support but is not itself supported. Thus the ground never falls but often stops a fall. Examples of kinds of support include resting on something, being fixed to something or hanging from something. Water and air can also support things (floating), this support often being partial.

To support something needs 'strength' or 'effort' (or both). Thus a shelf supports books by being strong; alternatively an aeroplane or a bird can support itself by its own effort, by flying. People support things (e.g. carry loads) using both strength and effort.

If the strength of a support is not enough, it may break. If the strength is enough the support takes (that is, absorbs) the weight of things it supports. We do not have to think of a well supported object having weight, unless the support is liable to break. As a support, the ground is infinitely strong and cannot break.

There can also be partial support. A swimmer may be partially supported by the water, and may make up the rest of the support by the effort of by swimming. A partial support means a partial fall, such as sinking. A dropped piece of tissue 'floats' down, partially supported by air.

A law of falling is that, having started to fall, things fall more rapidly the higher up they start and the heavier they are.

For these reasons, movement is conceptualised as taking place either on the ground (or on something supported by the ground), or as taking place in the air, above the ground. Motions which go up or down are distinguished from those which merely 'go along'. In this sense, the 'space' of motions has a preferred direction.

To describe motion further we need two more basic concepts, 'place' and 'path'. An object sitting still is at a certain place relative to other objects – on, under, beside, etc. One kind of motion consists of changing the place of something, as in passing a plate or pushing something aside. Another kind of motion is that in which the object is moving by itself – going on its way. The path it is following, and where it is along that path, is what locates it, not the place it happens to be in any moment. Motions are judged relative to the ground.

Motions of both kinds require effort, unless achieved by falling. Effort is used to change the place of something; to change the path, including starting and stopping; and to keep going along the path. Any lifting or raising involved requires additional effort. There are three possible sources of effort: effort of another agent on the object; effort generated by the object; effort of the present movement of the object.

Thus if you hand me a book or pass the salt you supply effort *on* the object to change its position. If you kick a ball along the ground you supply effort *on* the ball to start it going, after which it rolls using the effort *of* its motion. An athlete running or a car being driven use effort generated by themselves, in order to keep moving and, if they need to swerve or stop, to change path.

The effort needed is larger the heavier the object. The effort to start or keep moving is larger the larger the speed. If place is being changed, the effort is larger the larger the change of position; if path, the larger the effect on speed and direction of path.

The character of each kind of motion depends on the kind of support present. An object such as a bird or aeroplane uses effort both to support itself and to keep itself going. A ball thrown upwards in the air has effort *on* it from the thrower, but uses the effort *of* its motion to rise. When this is used up, since it has no effort to support itself and is not supported, it falls.

The effort of the motion has the special characteristic that it cannot be used to change the path of the same object (otherwise a motion would control itself). An object has no effort *of* motion when it is at rest relative to the ground.

The effort *of* motion is handed on dynamically moment by moment. The present motion makes the coming motion. When the speed changes little, as with a tennis ball or a dart, motion along the path is easy, with little or no effort being taken away from the effort of the motion. A motion like this uses up little effort, but still employs effort to keep going.

APPENDIX IV

BIBLIOGRAPHY

Andersson, B. (1990) Pupils' conception of matter and its transformation (age 12-16). *Studies in Science Education,* 18, pp58-85.

Arnold, P., Sarge, A. and Worrall, L. (1995) Children's knowledge of the Earth's shape and its gravitational field. *International Journal of Science Education* 17 (5) pp635-641.

Bar, V., Zinn, B., Goldmuntz, R. and Sneieder, C. (1994) Children's concepts about weight and free fall. *Science Education* 78 (2) pp149-169.

Bar, V., Zinn, B. and Rubin, E. (1997) Children's ideas about action at a distance. *International Journal of Science Education* Vol 19 no 10 pp1137-1157

Bliss, J., Ogborn, J. and Whitelock, D. (1989) Secondary school pupils' common sense theories of motion. *International Journal of Science Education* 11 (3) pp261 - 272.(12yrs-18yrs)

Bliss, J and Ogborn, J, (1993) A common-sense theory of motion. In Black P. J. and Lucas, A. *Children's informal ideas in science.* Routledge. London and New York.

Bliss, J. and Ogborn, J. (1994) Force and Motion from the Beginning. *Learning and Instruction* 4 pp7 - 25.

Brown, D.E. and Clement, J. (1987) 'Misconceptions concerning Newton's law of action and reaction: the underestimated importance of the third law', in Novak, J.D. (ed.), *Proceedings of the Second International Seminar: Misconceptions and Educational Strategies in Science and Mathematics* (Volume III), Ithaca, N.Y., pp39-53.

Brown, D.E. (1994) Facilitating conceptual change using analogies and explanatory models. *International Journal of Science Education* 16 (2) pp201 - 214.

Chaiklin, S. and Lave, J. (1993) *Understanding practice.* Cambridge University Press.

Champagne, A.B., Klopfer, L.E., Anderson, J.H. (1980) Factors influencing the learning of classical mechanics. *American journal of physics,* 48, pp1074-1079.

Clement, J. (1982) Students' preconceptions in introductory mechanics. *American Journal of Physics,*1 (50) pp66-71.

Clement, J., Brown, D.E. and Zietsman, A. (1989) Not all preconceptions are misconceptions: finding 'anchoring conceptions' for grounding instruction on students' intuitions. *International Journal of Science Education* 11 Special Issue pp554 - 565.

Dagher, Z.R. (1994) Does the Use of Analogies Contribute to Conceptual Change? *Science Education* 78 (6) pp601 - 614.

Di Sessa, A. (1983) Phenomenology and the evolution of intuition. In Mental Models Eds Gentner, D. and Stevens, A. pp15-33. Pub. Lawrence Erlbaum Associates.

Donaldson, M. (1978) Children's Minds. Fontana Collins.

Dreyfus, A., Jungwirth, E. and Eliovitch, R. (1990) Applying the "Cognitive Conflict" Strategy for Conceptual Change - Some Implications, Difficulties, and Problems. *Science Education* 74 (5) pp555-569.

Driver, R., Squires, A., Rushworth, P. and Wood-Robinson, P. (1994) *Making Sense of Secondary. Science Research into Children's Ideas*. Routledge.

Duit, R. (1991) On the Role of Analogies and Metaphors in Learning Science. *Science Education* 75 (6) pp649 - 672.

Dykstra, D. (1991) Studying conceptual change:constructing new understandings. In Research in physics learning:Theoretical issues and empirical studies. Eds Duit, Goldberg and Neidderer.

Dykstra, D.I., Boyle, C.F. and Monarch, I.A. (1992) Studying Conceptual Change in Learning Physics. *Science Education* 76 (6) pp615-652.

Eckstein, S.G. and Shemesh, M. (1989) Development of children's ideas on motion: intuition vs. logical thinking. *International Journal of Science Education* 11 (3) pp327 - 336.

Eckstein S.G. and Shemesh, M. (1993) Stage theory of the development of alternative conceptions. *Journal of Research in Science Teaching*, 30 pp45-64

Eckstein, S.G. and Kozhevnikov, M. (1997) Parallelism in the development of children's ideas and the historical development of projectile motion theories. *International Journal of Science Education* vol 19, no. 9 pp1040-1057 (grades 3 to 12)

Enderstein, L.G. and Spargo, P.E. (1996) Beliefs regarding force and motion: a longitudinal and cross-cultural study of South African school pupils. *International Journal of Science Education* 18 (4) pp479 - 492.

Engel-Clough, E. and Driver, R. (1986) A Study of Consistency in the Use of Students' Conceptual Frameworks Across Different Task Contexts. *Science Education* 70 (4) pp473 - 496.

Erikson, G. and Hobbs, E. (1978) 'The developmental study of student beliefs about force concepts', *Paper presented to the 1978 Annual Convention of the Canadian Society for the Study of Education*. 2 June, London, Ontario, Canada.

Finegold, M. and Gorsky, P. (1991) Students' concepts of force as applied to related physical systems: A search for consistency. *International Journal of Science Education* 13 (1) pp97-113.

Fischbein, E., Stavy, R. and Ma-Naim, H. (1989) The psychological structure of naive impetus conceptions. *International Journal of Science Education* 11 (1) pp71 - 81.

Gair, J. and Stancliffe, D.T. (1988) Talking about toys; an investigation of children's ideas about force and energy. *Research in Science and Technological Education* 6 (2) pp167-181.

Galili, I. and Bar, V. (1992) Motion implies force: where to expect vestiges of the misconception? *International Journal of Science Education* 14 (1) pp63 - 81.

Galili, I. (1993) Weight and gravity: teachers' ambiguity and students' confusion about the concepts. *International Journal of Science Education* 15 (2) pp149 - 162.

Galili, I. and Kaplan, D. (1996) Students' Operations with the Weight Concept. *Science Education* 80 (4) pp457-487.

Gilbert, J.K., Watts, D.M. and Osborne, R.J. (1982) Students' conceptions of ideas in mechanics. *Physics Education* 17 pp62-66.

Gilbert, J.K. and Watts, D.M.(1983) Misconceptions and alternative conceptions: Changing perspectives in science education. *Studies in Science Education*. 10, pp61-98.

Gilbert, J.K. and Zylbersztajn, A.(1985) A conceptual framework for science education: The case study of force and movement. *European Journal of Science Education*. 7(2) pp107-120/

Graham, T. and Berry, J. (1996) A hierarchical model of the development of student understanding of momentum. *International Journal of Science Education* 18 (1) pp75 - 89.

Gunstone, R.F., Gray, C.M.R. and Searle, P. (1992) Some Long-Term Effects of Uninformed Conceptual Change. *Science Education* 76 (2) pp175 - 197.

Gutierrez, R. and Ogborn, J. (1992) A causal framework for analysing alternative conceptions. *International Journal of Science Education* 14 (2) pp201 - 220.

Halloun, I.A. and Hestenes, D. (1985) The initial knowledge state of college physics students. *American Journal of Physics* 53 (11) pp1043-1055.

Hamilton, D.J. (1996) The peer interview about complex events: a new method for the investigation of preinstructional knowledge. *International Journal of Science Education* 18 (4) pp493 - 506.

Hatano, G. and Inagaki, K. (1992) Desituating cognition through the construction of Conceptual Knowledge in Light, P. and Butterworth, G. (Eds) Context and cognition, London Harvester.

Hennessy, S. (1993) Situated Cognition and Cognitive Apprenticeship: Implications for Classroom Learning. *Studies in Science Education* 22 pp1 - 41.

Hermann, F. and Schmid, G B (1984) Statics in the momentum current picture. *American Journal of Physics*. Vol. 52, No.2, Feb.

Howe, A. (1996) Development of Science Concepts within a Vygotskian Framework. *Science Education* 80 (1) pp35-51.

Karmiloff-Smith, A. (1992) *Beyond modularity*. A Developmental Perspective on Cognitive Science. *MIT Press*.

Kruger, C., Summers, M. (1990) A survey of primary school teachers' conceptions of force and motion. *Educational Research* 32 pp283-294.

Kuhn, D., Garcia-Mila, M., Zohar, A. and Andersen, (1995) Strategies of knowledge Acquisition. *Monographs of the society for research in child development* no 245 Vol 60 no 4. University of Chicago Press.

Kuiper, J. (1994) Student ideas of science concepts: alternative frameworks? *International Journal of Science Education* 16 (3) pp279 - 292.

Lave, J. and Wenger, E. (1995) *Situated Learning*; *Legitimate Peripheral Participation*. Cambridge University Press.

Laurendau, M. and Pinard, A. (1962) Causal thinking in the child:A genetic and experimental approach. New York:International Universities Press.

Lee, K and Karmiloff-Smith, A. (1996) The development of external symbol systems: The child as notator. In Gelman R and Kit-Fong Au, T (1996) Perceptual and cognitive development. Academic Press.

Lemeignan, G. and Weil-Barais, A. (1994) A developmental approach to cognitive change in mechanics. *International Journal of Science Education* 16 (1) pp99 - 120.

Lythcott, J,. and Duschl, R. (1990) Qualitative Research: From Methods to Conclusions. *Science Education* (74) (4) pp445-460.

Mariani, M.C. and Ogborn, J. (1991) Towards an ontology of common-sense reasoning. *International Journal of Science Education* 13 (1) pp69-85

McClelland J.A.G. (1985) Misconceptions in mechanics and how to avoid them. *Physics Education* 20 pp159-162 G6,P,M.

McCloskey, M., Caramazza, A. and Green, B. (1980). Curvilinear motion in the absence of external forces: Naive beliefs about the motion of objects. *Science 210* pp1139-1141.

McCloskey, M. (1983a). Intuitive Physics. *Scientific American* 248 (4) pp114 - 122 g6,P,M.

McCloskey, M. (1983b). Naive theories of motion. Gentner, D., Stevens, A.L.: *Mental models*. Hillsdale and London: Lawrence Erlbaum pp299 -324.

McCloskey, M. and Washburn, A. and Felch, R. (1983) Intuitive physics: the straight down belief and its origin. *Journal of Experimental Psychology: Learning, Memory and Cognition, 9* pp636-649.

McCloskcy, M., Kohl, D. (1983). Naive physics: the curvilinear impetus principle and its role in interactions with moving objects. *Journal of Experimental Psychology* 9 pp146 - 156.

McCloskey, M. and Kargon, R. (1988) The meaning and use of historical models in the study of intuitive physics.pp49-67. In *Ontogeny, phylogeny and historical development* Ed. Strauss, S. Pub. Ablex

McDermott, L.C. (1984) Research on conceptual understanding in mechanics. *Physics Today* 37 (6) pp24 - 32.

McGuigan, L. and Russell, T. (1997) A grounded theory model of constructivist practice. Paper presented to the *First Conference of the European Science Education Research Association,* Rome. September 1997.

Millar, R. and Kragh, W. (1994) Alternative frameworks or context-specific reasoning? Children's ideas about the motion of projectiles. *School Science Review 75,* 272 pp27-34.

Minstrell, J. (1982) Explaining the "at rest" condition of an object. *The Physics Teacher 20.* pp10-14.

Mullet, E. (1990) Distinction between the concepts of weight and mass in high school students. *International Journal of Science Education* 12 (2) pp217 - 226.

Ogborn, J. (1985) Understanding students' understandings: An example from dynamics. *European Journal science Education* 7 (2) pp141-150.

Osborne, R. (1980) Force Learning in Science Project. Working Paper No 16. University of Waikato. Hamilton NZ.

Osborne, R. and Gilbert, J.A. (1980) A Method for the Investigation of Concept Understanding in Science. *European Journal of Science Education* 26 (3) pp311-321.

Osborne, R., Schollum, B. and Hill, G. (1981) Towards changing children's ideas. Force, Friction and Gravity - Notes for Teachers. Learning in Science Project, University of Waikato.

Pfundt, H. and Duit, R. (1997) Bibliography Students' Alternative Frameworks and Science Education Kiel IPN.

Piaget, J. (1929) The child's conception of the world. New York. Harcourt Brace.

Palmer, D (1997) The effect of context on students' reasoning about forces. International journal of science education Vol 19 no6.

Reif, F. (1987) Instructional design, cognition and technology: Applications to the teaching of scientific concepts. *Journal of Research in Science Teaching*, 24 pp309-324.

Reynoso, E.H., Fierro, E.H. and Torres, G.O., Vinventi-Missoni, M. and Perez, J. (1993) The alternative frameworks presented by Mexican students and teachers concerning the free fall of bodies. *International Journal of Science Education* 15 (2) pp127 - 138.

Ruggiero, S., Cartelli, A. Dupre, F. and Vincentini-Missoni,M. (1985) Weight, gravity and air pressure:Mental representations by Italian middle school pupils. *European Journal of Science Education* 7 (2) pp181-194.

Russell, T., Qualter, A., McGuigan, L. (1994) Evaluation of the implementaion of Science in the National Curriculum at key Stages 1, 2 and 3. Volume II Progression Schools Curriculum and Assessment Authority.

Schilling, M., Atkinson, H., Boyes, E., Qualter, A. and Russell, T. (1992) Knowledge and Understanding of Science. Forces. A Guide for Teachers. National Curriculum Council.

Sequiera, M. and Leite, L. (1991) Alternative conceptions and history of science in physics teacher education. *Science Education*, 75 91 pp45-56.

Schollum, B., Hill, G. and Osborne, R. (eds.) (1981) Towards changing children's ideas Teaching About Force - Suggested Teaching Activities. Learning in Science Project, University of Waikato.

Scott, P. Asoko, H.M. and Driver, R. (1992) Teaching for conceptual change:A review of strategies. Institut die padagogik der naturwissenschaften an der. In Research in physics Learning: Theoretical Issues and Empirical Studies. Eds Duit, R, Goldberg, F. and Neidderer, H.

Sharratt, M. (1994) Galileo. Decisive innovator. Blackwell, Oxford UK and Cambridge USA.

Shulman,L. (1987) Knowledge and teaching: foundations of the new reform. *Harvard Educational Review,* vol 57 no1 pp122.

Simon, S., Black P. and Brown, M. (1994) Progression in understanding forces: metacognitive reflections. *Paper presented at the Annual Conference of the British Educational Research Association: Oxford.*

Solomon, J. (1993) The social construction of children's scientific knowledge. In, Black P. J. and Lucas, A. Children's informal ideas in science. Routledge. London and New York.

Stcad, K. and Osborne, R. (1980) *Friction,* LISP Working Paper 19, Science Education Research Unit, University of Waikato, Hamilton, New Zealand.

Stead, K. and Osborne, R. (1980) *Gravity,* LISP Working Paper 20, Science Education Research Unit, University of Waikato, Hamilton, New Zealand.

Stead, K.E. and Osborne, R.J. (1981) 'What is friction: some children's ideas', *New Zealand Science Teacher* 27 pp51-57.

Steinberg, M.S., Brown, D.E. and Clement, J. (1990) Genius is not immune to persistent misconceptions: conceptual difficulties impeding Isaac Newton and contemporary physics students. *International Journal of Science Education* 12 (3) pp265 - 273.

Terry, C., Jones, G. and Hurford, W. (1985) 'Children's conceptual understanding of force and equilibrium', *Physics Education* 20 (4) :pp162-165.

Thijs, G.D. (1992) Evaluation of an Introductory Course on "Force" Considering Students' Preconceptions. *Science Education* 76 (2) pp155 - 174.

Thijs, G.D. and Van Den Berg, E. (1994) Cultural factors in the origin and remediation of alternative conceptions in physics. *Science and Education* 4 (4) pp1-32.

Thijs, G.D. and Bosch, G.M. (1995) Cognitive effects of science experiments focusing on students' preconceptions of force: a comparison of demonstrations and small-group practicals. *International Journal of Science Education* 17 (3) pp311 - 323.

Treagust, D.F. Duit, R., Joslin, P,. and Lindauer, I. (1992) Science teachers' use of analogies: observations from classroom practice. *International Journal of Science Education* 14 (4) pp413 - 422.

Twigger, D., Byard, M., Driver, R., Draper, S., Hartley, R., Hennessy, S., Mohamed, R., O'Malley, C., O'Shea, T. and Scanlon, E. (1994) The conception of force and motion of students aged between 10 and 15 years: an interview study designed to guide instruction. *International Journal of Science Education* 16 (2) pp215 - 229.

Viennot, L. (1979) Spontaneous Reasoning in Elementary Dynamics. *International Journal of Science Education* 2 (1) pp205 - 221.

Vosniadou, S. and Brewer, W.F. (1994) Mental Models of the Day/Night Cycle. *Cognitive Science* 18 pp123 - 183.

Vygotsky, L.S. (1978) Mind in Society. The Development of Higher Psychological Processes. Harvard University Press.

Wason, P.C. (1977) The theory of formal operations- a critique. In Piaget and knowing. Ed Gerber, B.A. Routledge

Watts, D.M. and Zylbersztajn, A. (1981) A survey of some children's ideas about force. *Physics Education* 6 (16) pp360-365.

Watts, D.M. (1982) Gravity - don't take it for granted! *Physics Education*. 17 (4) pp116-121.

Watts, D.M (1983) A study of school children's alternative frameworks of the concept of force. *European Journal of Science Education,* 5 (2) pp217-230

White, R. and Mitchell, I.J. (1994) Metacognition and the Quality of Learning. *Studies in Science Education* 23 pp21 - 37.

Whitelock, D. (1991) Investigating a Model of Common Sense Thinking about Causes of Motion with 7 -16 Year Olds. *International Journal of Science Education*. 13 (3) pp321-340.